# The Delivery Man

*The Art of turning ideas into
products in Silicon Valley*

*By Sebastien Taveau*

*To my parents, Jean and Michèle,*
*To my children, Aleksandre and Rachelle,*
*To my wife, Rebekah,*

*Your unconditional support and love through my professional trials and tribulations, even when it affected our personal life, is a gift I cannot repay.*

# Table of Contents

# Foreword by Dave Birch

First of all, let me disclose an interest: I have known Seb for many years and have watched him navigate the treacherous waters of Silicon Valley through startups, scale-ups and screw ups. He has been an invaluable barometer showing me what is coming next, from non-bank payments to biometric authentication to embedded finance and APIs.

Hence, I took great delight in reading his memoir-cum-manual about product delivery in exciting times, because the ability to actually deliver a new product, to go from the idea sketched out on a napkin to the sustainable revenue stream and satisfaction for all stakeholders, is essential to the economy but (as Seb himself notes) unglamorous.

We focus on the front men and women who generate the good and bad ideas, but we don't always see that behind them are people who actually make things happen, the unsung heroes of product development. Seb's career has spanned the transition from waterfalls to agile, through rapid prototyping and, dare I say it, tinkering.

If there is a central thread, it is about security. In an online and interconnected world, a deep and abiding appreciation for risk analysis, countermeasures and cybersecurity is necessary to create practical scale solutions. You simply cannot cut corners on this when you are delivering products that must work for Zelle, for Mastercard, for PayPal and others.

While many of the vignettes in the book are helpful lessons in themselves, they come together to show a trajectory worth emulating. Missions, managers, and mentors are the components of a career creating products and services that made a difference despite unstable technology platforms and even more unstable CEOs.

How do you do this and stay sane? There are toolkits, such as the situational principle that Seb discusses here, but my inexpert observation is that people play the key role and from my experience working with venture capitalists (including as a judge for one of Seb's hackathons!) it seems to me that the ability to assemble and direct a team that can deliver is central to the Valley's momentum. This is no longer a world of heroic 10x programming but a world of co-ordination and trust, something that Seb has as fundamental characteristic. Not all of us are Elon Musk. Instead, finding how to, as Seb says, federate talents is in practice the way to change the world!

This memoir is, I hope, a useful addition to the toolkit for those entering Silicon Valley and presents a plain but important message: If you can deliver, you can build a wonderful career.

*David Birch - Author, Advisor and Commentator on Digital Finance.*

# Introduction

*"Imagine" - John Lennon*

The idea of going against one of the most ingrained myths of Silicon Valley has always been at the bottom of my list, but here we are. You don't need to be the greatest innovator to make a living in the mecca of technology.

Everything in this book is based on my experience. How did the idea come to life? I was on stage in Brazil talking to an audience of bank executives, fintech entrepreneurs and young students, all waiting to listen to the wise words of a Silicon Valley innovator. Yep, me.

As I was prepping for the talk, I looked into my career path. What led me to that specific spot where people would take advice from me because of where

I work? Was my own experience even relevant? The fascination with Silicon Valley for people outside of the US is sometimes borderline cultist. In Silicon Valley, techies' entrepreneurs also like listening to themselves talking. So why not make a book out of it?

For the audience, the shockwave I sent after a few seconds was not what they were expecting. I told them I was a fraud, not an innovator, just a guy doing his job with no glamor nor spotlights, delivering what I was supposed to put together. I was an integrator of technological concepts and product requirements.

This led to the conversation around the magic "I"s and who we were instead of who we were or wanted to be perceived to be.

So let me explain the context around these "I's." Initially, there were only four, then over time, a few were added, like Idea, Implementor, Instigator, Investor, Incubator, etc.

former bosses were leaning more on the integration side. Few had the mind of an innovator, and one definitely had the mind of an inventor, but it reaffirmed that despite having worked on major projects, it was mostly the timing of the idea implementation determining if it was innovative or just better done than the others.

So, who am I to be dispensing unwanted but clarifying advice on a career which looks stellar on LinkedIn but got stories behind each role?

Well, here is what I have delivered or was closely a part of the core team working on:

- PayPal Mobile: First mobile apps incl. mobile security and identity in cross-payments
- Mastercard Masters of Code: First global developer event series for a network with OpenAPI
- Zelle: First real-time payment network in North America.
- Validity: First consumer biometrics solution for mobile devices (yep, before Apple TouchID)

- Envestnet | Yodlee: First full-fledged developer program to support new Financial Wellness and Wealthtech ecosystems
- Metrowerks / Codewarrior: First true developer suite, including the first compiler for new PPC microchips used in Apple computers back in the 1990s, which saved them because no apps would have shipped on it otherwise in the first few months
- Renesas/Hitachi: First fully certified e-Payment reference terminal

So if you are a fresh graduate or a young product/program manager or a business analyst trying to figure out how to do things practically or if you are an executive trying to understand the best way to guide your team, this book will be a nice addition to many theories you may have learned or heard about.

Everyone wants to be an innovator, yet you can be successful in being the delivery person for a visionary (or you can be the visionary).

A visionary without a delivery person has an increased level of risk of failing. Silicon Valley is the graveyard of many ideas that never reached the market because of timing and delivery.

So now, while the Delivery Man can be quite unnoticed within a company, the role is critical to the success of many product launches. It is important to have a basic understanding of how products come to existence, especially in Silicon Valley technology companies. The basics are simple.

> There is a need in the market, either expressed by consumer demand or market research. The leadership of a company recognizes this need, and it is decided to have a strategy to address the need.

> The strategy is in place. Business requirements come into existence on basic expectations.

> A project is put together (the project is the short-term process to get the product out while

the product is the permanent outcome) and the budget and resources assigned to it.

An executive sponsor must approve the project before it starts to support the overall strategy. Otherwise, you will get in big trouble.

A business analyst is usually in charge of these requirements, including the main scope.

A product manager (who sometimes is the business analyst in small companies) takes on these business requirements and turns them into product requirements.

Meanwhile, a project manager also joins the crew to help put together the timeline, resource allocation, risk management, and detailed scope framing (the grooming sessions). To be noted that a program manager is a person who manages multiple projects needed to work together to deliver the overall intended experience. Programs are also more strategic than a short-lived punctual project, meaning a program may cover the same length of time as a product life.

The project is started, the product manager makes sure the development team understands the requirements, the project manager makes sure the project remains on track and no additional requirements come in (best case scenario, but it never happens). The executive sponsor is kept abreast of the progress, and product ongoing deliveries are approved by the product manager and business analyst (the demo sessions).

When everything is agreed on and properly tested, the product is released to production (from development) and for the world to enjoy. The above is the simplified theory of how things should work to complete a delivery. But of course, there is often a wide-open gap between theory and practical reality.

Let's bridge the gap through key lessons learnt from concrete hands-on experience.

# CHAPTER 1
# Becoming a delivery man

*"Immigrant Song" - Led Zeppelin*

As soon as I landed for the first time in the US, Silicon Valley was always the top reference for everything tech, even when I was working in Austin, TX. As a young immigrant mixing tech and business, I fancied myself becoming one of these zillionaires' techies out of Silicon Valley. The dream of riches and fame was alluring. It didn't hurt that I had met Jean-Louis Gassée as a young man in France and seen his stellar success as a Silicon Valley Baron.

I started my life in the US as a young intern under a visa J-1 in the mid-90s. It was a time of rapid growth in the world of technology, and I was so happy to just be in the US and earning a whopping $16,000 per year. It was supposed to be for six months, but after that period, the company I worked for decided to extend my stay by converting my visa to the magic H1-B and gave me a massive raise as a full-time employee granting me the earning of $20,000 per year. Yes, it is ironic now that I know better, but I was living my dream. The first few years of my career were spent in companies either having a product already in place or being big enough that it was hard to pull the "innovator" hat most of the time. Then, finally, I jumped into the world of start-ups hoping to learn the skills to be great at tech and a success story when the start-up would be massively successful, and the financial exit completed (Hint: it never happened).

In the world of start-ups, I was good at finding companies working on super advanced technologies but always 10 to 15 years too early. For example, I worked for a company subsidized by Texas Instruments on their first full mobile platform (OMAP). The core operating system was DSPLinux which later became the core foundation of Android. But again, too early in the game.

Talking about games, I also worked for a gaming company that invented the first Massive Multi-Player engine to generate graphics and visuals in real-time for games requiring scenes with many characters. Technically, we leveraged AI (artificial intelligence) before AI was even cool. Gaming has always been quite in advance from a technology viewpoint, but we were way too advanced on this one.

This led to a critical juncture in my life. Few people know about it, but it shaped how I approached a lot of my work decisions, even now. In 2002, without a job after the 2001 collapse of the tech bubble and post 9-11, with a newborn (with medical complications during the delivery) and the semiconductor industry in Austin being devastated, my wife and I had to decide our next move.

Almost penniless, on the verge of not making the next mortgage payment, we sold our house and moved to the Bay Area, where my wife was originally from. We had met in Texas, where she was completing her Ph.D. at the University of Texas, but she was truly a bi-coastal girl. So, when we had to pick, it was Montreal or Boston on the East Coast to be close to my family or California to be close to her family. And California won.

Before moving, I drove from Texas to the Bay Area with my pick-up truck and my dog Sinclair in tow. I had six job interviews lined up, and it seemed things would improve. Most people were leaving the Bay Area also devastated by the bubble burst, and few were coming in (which seems to be a regular cycle of the Silicon Valley pulse). I took a job in a start-up in the Bay Area, and it didn't go as planned. While my mind was more on how I should have negotiated more stock options, I didn't pay attention to the core financials, and I learned the hard way when salaries were paid whenever the money could be there. It was upsetting especially when we were coming out of a stressful slice of life (marriage, relocation, birth, and finance are the biggest tension points in a couple's life, and we did all of them almost simultaneously). However, a couple of years later, while chatting with someone who knew the founder, he told me something that I have since profoundly admired. To be able to pay his employees' salaries, and as irregularly these may have been, the founder and CEO had refinanced his house and taken on a personal loan. He had put his family's financial well-being in jeopardy to pursue his dream but also to ensure his employees would be taken care of. He knew his purpose.

And now, despite setbacks, we were finally in the heart of Silicon Valley, ready to live the dream.

The mobile revolution was starting, "vibing" all around. Everyone was focused on size, capacity, speed, connectivity, etc. All good but one of the key elements of mobile devices was the lack of a true security approach. Components were all working together, but memory leakage was common. But again, these were not smart devices but mostly featured phones. This didn't mean security was unimportant, and few companies worked on it at different stack levels. I joined one of these, which was a spin-off from a large European smart card manufacturer. The core element in all devices was the SIM card and the company had developed a way to manage a certain level of security via software vs additional costly hardware components. And the first segment to go after was payment. Again, too early. People were still transitioning to the fact you could have a mobile device to make phone calls, keep an address book with your contacts and sometimes even have a digital copy of your agenda. That was the mindset. Payment was not even close to getting on the radar of main street for another ten years. Even if Japan was exploring these, the rest of the world was not ready.

Now again on the lookout for a new job, I made the big jump and became a true entrepreneur. I started my consulting company, Fluctuat LLC, in 2004. The idea to have a business dated back to 1998 while in Texas, but I never had the courage to pull the pin. The clever name (or so I thought) was a take on Fluctuate, for the fluctuations of the markets, the fluctuations of the silicon components and Fluctuat Nec Mergitur, the motto for the City of Paris in France, where I attended the University of Paris 1 – Panthéon Sorbonne.

As we expanded the business, I added more customers to my portfolio. Being an entrepreneur is difficult. It meant doing the job we were paid for during the day, solving problems in the evening and doing all the administrative work and prospect alignments on weekends (like answering RFPs). The clever idea of Fluctuat LLC was to be an umbrella of independent consultants/contractors with various expertise who could be engaged via this "single entity" carrying insurances, infrastructure cost, taxes, etc. We were all 1099s from a tax perspective as there was no employee. Looking back, it was like the Uber of tech consultancy.

The nice perk of the job was to be brought in on some innovative projects and have the capacity to have an impact being "the external expert." I could now say I was running a tech business in Silicon Valley, had achieved the title of CEO before turning 40 and was exposed to some of the coolest ideas and people I had met before. It looked like ideas and start-ups were popping left and right wherever you were going. My tech brain was overstimulated by all the possibilities but enjoying it. Consulting for a few Japanese companies allowed me to acquire a unique understanding of business, technology and strategy as Japan was more advanced than most countries on the mobile front. My unique background was getting me to run around the world and understand how technology was applied in Asia, Europe, North and South America, Australia, etc. It was exhilarating to be part of what was happening around mobile devices, phones, PDAs, tablets, etc.

It was great while it lasted, but a couple of "partners" I added in France took off with the concept and cut the US side of the business from deals. While dividends were supposed to be paid, they started "forgetting" to report deals or incomes. This became untenable, and I completely removed the US side of the business from the deal. It cost me on a financial and personal level quite a lot. It was the crashing sound of my American Dream.

But when you are at the bottom, it also gives you a perspective of what do you want to do next? Keep sitting at the bottom of the hole and feeling bad about your situation or climbing out and trying something new?

The rebound for me came with PayPal. The new executive for the Mobile team chatted with me informally during an event, and we realized we knew each other from a previous professional life. He was the head of a business unit I was prospecting at the time for Fluctuat, but which ended up passing on a contract. It was not a big deal as we signed with the competition right after.

The chance I was given at PayPal came via Eric D. And for this, I am grateful. It put my career on a path I was not seeing. Larger than life, Eric, the newly hired head of PayPal mobile, was this super tall person from Gascony in France. If you never read the Three Musketeers, you are missing the eccentric panache associated with people from this region. As we discussed his new role at PayPal, he invited me on campus to chat more. He was working on the mobile roadmap for the PayPal Mobile unit and was interested in my opinion on contactless technologies beyond the usual carrier signals (as in Bluetooth and NFC). Since it was one of my areas of expertise working with electronic component companies (and my time at Motorola), he offered me a role in his team. I was to work on mobile technology integration and strategy. It was what looked like a perfect fit. Fun fact, I thought he was negotiating a consulting gig with me when he was hiring me out.

To this day, I can partially claim that PayPal bought me out of my company. But it never bought my company, which was disappointing. I'll take it as a 50% win which is ain't bad in Silicon Valley.

Anyway, going back to PayPal. The best thing happened to me. We were Generation 2. We were building the mobile revolution. No one knew how to approach it. The iPhone had been announced but not released yet, the switch from feature phones to smartphones was not accomplished yet, and tablets were still supposed to be subpar computers running on the same operating system.

Among these conflicting technologies, chaotic companies, and confusing products, it seemed like PayPal was fiercely efficient at weeding out superfluous concepts. With a similar background, Eric D. and I focused on the hardware security part of the equation, which led to our group generating more than fifty percent of all the patents for PayPal over a two-year period. A treasure chest with a lot of value today with deployment and features coming from other companies but covered by such patents.

You cannot beat these early days from a defensive intellectual property viewpoint.

Until my time at PayPal as indicated before, I was good at finding companies working on super advanced technologies but always ten to fifteen years too early. At PayPal, I learnt to focus on much shorter innovation, affecting what was a long-term vision of five years and optimizing delivery roadmaps to palatable milestones of six months, one-year, and two-year goal lines. It did help that I was also surrounded by very talented business, product, project, engineering, and marketing people. Many of these former colleagues have moved on to greater things, becoming founders, venture capitalists, executives in large firms and more. The nice part is that most of us are still in contact and regularly meet at events to exchange stories on success, job situations, challenges, remember the good old days, or cherish the memories of some who are no longer with us. Sharing the experience and the same training and work skills with a group of people who can understand you without question is an invaluable tool to reflect on who you are.

While my role at PayPal was still around "future" technologies, these were now grounded in practicality and deliverability in support of well-defined and timed roadmaps.

There, with one of the best titles I ever held, "Astronomer," which resulted from a bet made many years before on the Arctic Circle while visiting Nokia and their famous scientist Timo T. (if Nokia phones were so robust, it was because of the T^3 process, Timo T. Test). Long story short, I won the bet, and if you buy a copy of these books and we meet, I will sign it for you and tell you the details around a drink.

As Ze Astronomer at PayPal, my role was to "keep an eye on all the stars in the sky and tell the captain of the boat where to go." In plain English, I was surveying the technology and standards landscape and ensuring that PayPal had a say or play early enough in the process. We could integrate, build or buy to support the mobile strategy if it made sense. We wanted to be front and center for payments in the new mobile ecosystem being built.

The experience acquired became valuable in future roles. With the capacity to "see" what needs to be delivered and when to satisfy the leadership team, the shareholders, the team aspiration and, more importantly, the consumer needs, it grants the capacity to operate with more autonomy and freedom. Being self-sufficient on the top is an even more welcome skill to nurture. When everything seems to be going well and being executed and delivered with little fanfare, it will keep away micro-managers. They would be lost in a world of consistent deliveries despite some unknown (more of this in Chapter 3). Having the gift of seeing the finished product, whatever it is and all its components and moving parts gives the capacity to react and have a flexible roadmap to a point without running into more troubles.

This autonomy has allowed me to work on projects that became very visible, but the experience of being a better delivery person (and timekeeper) acquired at PayPal set me up for it.

I would not have worked on the Mastercard Masters of Code program and the OpenAPI developer portal built to expose Mastercard technology to the wide world. The freedom to execute the vision and the way to shape the Masters of Code was a labor of love. It became one of the biggest developer series in the world. While some internal initiatives were partially competing in some aspects, the capacity to create a coalition of goodwill employees and partners, even coming from these other initiatives, helped create this unique moment in Mastercard and the developer community when everything seemed in unison. The reason is that I had learnt to articulate what we will deliver, when and what it will look like to my boss and the executives in various geographies. This created a shared vision, a common enthusiasm, and a larger budget.

After we delivered the first hackathon of the series, the news spread around the company. It helped to have Anne Cairns, president of the international business for Mastercard as our first executive judge. While it looked easy, there were a lot of small hiccups. It didn't derail the final delivery of the event, but we integrated these to make the next events better for everyone. We learnt from them and "adapted" our playbook/roadmap to make sure we could deliver with more efficiency.

What it did, as we moved further into the global series of hackathons, was to give us more buzz. As we reached the Grand Finale in San Francisco, many tech personalities were vested in attending. This was the culmination of a year-long journey, and we had to deliver the ultimate event. Again, with planning and experience, the vision of what it was supposed to look like up to some silly little details was planned and executed. To this day, that weekend still looks like a dream. It was unbelievable to have staff, developers from around the world, tech influencers, and a Hollywood celebrity converging on that single event we planned all year. And we delivered.

Fast forward to recent days, and that playbook I started my journey on at PayPal, refined at Mastercard, is still one of the core references I go back to when starting a new project. Many signs of it can be found in the launch of Zelle or the developer experience revamp at Yodlee and its supersizing at Envestnet. And it is not just unique to my experience. As mentioned earlier, the former colleagues at PayPal have also been using similar approaches on some of their projects. Going to the same great school left its mark ten or fifteen years later.

Being able to understand to the core what the expectations of the executive team are, the way to inform them of the progress and obstacles (avoid these as much as possible), and the capacity to align requirements with reality (we had a colleague once who requested companies answering an RFP to explain how they would support a security technology which didn't exist at the time. The standards were still being worked on, and it would not be until four years later that they would be finalized. That requirement was whacked from the RFP faster than you can hear the whacking sound and ensured the team is in full alignment on priorities and delivery sequencing. As the saying goes, do not put the cart before the horse.

I always fancied myself becoming an innovator in the cliche type of Steve Jobs. When I finally accepted it was silly and that pursuing this dream would make my life miserable, I discovered my true superpower. Understanding the mind of the visionary, converting it into a practical reality, and then delivering it to the world. Do I feel like I gave up on my original dream? No. I just reframed it. With lesser aggravations for my health and family. Win-Win-Win. I still work on advanced projects or innovative approaches to problems, but my role is to ensure the final integration of all the pieces will be delivered in good order. This is a complementary profile to any visionary or innovator who may not know how to convert their idea into practical implementation and/or to address a real market need.

**Key Takeaway:**
Learn what you are good at as soon as possible in your career and accept it even if it may not lead to your original dream.

# CHAPTER 2
# Finding unity in many

*"What a wonderful world" - Louis Armstrong*

This is not a one-person job, even if you are a great delivery person. You need the help and support of so many roles to get anything out of the door. It is important to quickly understand the strengths and motivations of each person willing to help on a project (don't dwell on the shortcomings). A great team delivers. The synchronization between individuals is the biggest challenge. Providing a positive outcome for people to rally behind your project will be your biggest challenge, proving you are a great delivery person.

In my experience, ensure you have the right people to cover all parts when the delivery process is well understood (if there is a process, as sometimes it is only partial or non-existent). Nothing starts without the executive vision or sponsoring. Don't waste time executing something that is not part of the executive roadmap. You may find passionate people to help, but it rarely brings any positive outcome. Usually, an executive mandate is supported by a strategic goal or a market need. And while it may seem ludicrous to provide advice from the perspective of large companies with large budgets and unlimited staff, I can guarantee you it worked for even smaller companies or being a single contributor. If the founder of a start-up is not on board, you won't have any voice to support the plan at the board of directors' level. If you are in doubt about the level of support, it probably means there is no support. From there, you will need to have a strong project management person (if it is not you) to keep all the pieces falling in place at the right time, then someone who understands product requirements very well and as it is in my field, someone who can interpret these requirements for the engineering tech team. Oh, and don't forget the testing/QA part. Don't wait for the end of a rollout to figure out testing was not done properly.

The best way to understand how to be the best manager to a team, where often people will not report to you directly but representatives of their own respective team, is mostly in a tale that my manager at PayPal told me once. It stuck with me up to this day.

You have three kinds of colleagues and managers. The Good, the Bad and the Ugly Donkey. At any time in your career, you will be or will face one of them. Be ready. The advice, solicited or not, was received at certain points in my career. They either motivated me to become a better person or completely turned me off from any energy spent on projects or companies. However, understanding how managers behave will help "articulate" and "communicate" better with teams assembled for a common goal of delivering something to the market.

Having seen the profiles below in all my colleagues, it has been one of my constant goals to work for managers or with colleagues fitting in the non-toxic category. However, sometimes, you must also try to understand why a team member or your boss may behave a certain way. If it is out of the ordinary, something is happening either at work or home. Giving some space is an option or sometimes just asking the simple question "do you need help with anything?" will get you a long way.

## *The GOOD*

What makes a great manager? For me, someone is attuned to their team and can say the right words at the right time and in the right context to improve a situation. Sometimes, these nuggets of wisdom will remain with you for a long time. So here are some of the advice/kind words I received over the years from my bosses.

- Praise in public, admonish in private

- Technology is like a diet. It's easier to gain weight than to lose it, so design it simply.

- Be transparent, and share information. People are more willing to help when they feel part of the team.

- Experience is what you get when you don't get what you want. So work harder to get what you want.

- "Great meeting. I learnt something." When a boss tells you that, you not just feel valued, you feel empowered to do more using this skill.

- "Where do you want to be in five years? Build a career plan and make it happen. I'll support you." This came late in the game for me, but it also happened when I was at PayPal. This came from a new manager for the team, Laura C. She had been working on roles with increasing responsibilities at eBay and PayPal. Even when things were not going her way, she always seemed able to stay focused and make the change needed for her to move forward. She is today a successful CEO of a company unrelated to the world of payment or mobile technology but making a bigger difference. That small piece of advice during a review put me on track to share the same advice with people reporting to me. If they can move up and succeed, within or without my team is not important. If they wanted to grow and I could help, why not?

On top of that, people in positions of power, top executives will sometimes agree to either mentor you or give you a 15-minute meeting which becomes a 1h meeting or, even better, a lifelong advisor. This is sometimes the most valuable gift executives, and bosses can give to their employees. I have been lucky enough to have a few of these mentors, and even up to this day, I am still close with some former "good" bosses and ask for neutral opinions to solve some challenges I may encounter.

## The BAD

But for all the positives, I also witnessed the worst. Not all bosses are cut to be people managers. Some should not be around people at all. So here are the worst comments/advices I heard, which are still the bottom of the jokes with former colleagues.

- *"Something must be wrong, the internets are not working."* While you may not be an expert at everything, if you are in a tech company, it is important to at least understand the basic technology concepts running the world today.

- *"Why do you want a raise? You should be happy to have a job."* This is the best way to have someone resign. Maybe it's part of the plan, but it is also very demoralizing to the rest of the team when morale is low across the board. As the latest buzzword says, you will have a team of ghost quitters to deal with, and nothing will be delivered on time or at the quality level expected.

- *"So the agenda of the meeting today has been sent earlier (it has not). Seb, can you lead the meeting and contribute?"* (*being put on the spot*) This is the opposite of a boss who shares information and helps plan meetings or agendas. You can't wing a meeting when you are the boss. Otherwise, you are wasting time for everyone.

- *"Well, I can't really give you an outstanding review because I really thought you should have volunteered to take that project over"* - Even when you take over a new team, and it is tempting to rely on people you trust and worked with in the past, evaluate the talents and see if they are used properly. She was a new boss and I was in tech. The project was a marketing/analytics project that was way outside of my competency. This cost me a great evaluation one month before the end of the year when all my previous bosses – had three that year – all gave me outstanding. However, being in her shoes today, besides having financial pressure to save money against the P&L, I would have hoped to approach the situation better. Being proactive is important, and as I noticed being involved in projects outside of my perceived expertise, I should have asked more questions. What is expected of me? How can I help the team and the manager? What is urgently needed that I may not have in my job description, but I may have some knowledge of and the bandwidth to provide help?

- *"I am the boss, so you do as I say."* Or *"This is not a democracy. There is no more discussion (there was none). You work for me so do as I said."* I especially love these because we have all heard them at least once in our career. This is the fastest way for a manager to ensure their authority is established and not challenged. It is also the fastest way to NOT deliver anything as you will have a highly demotivated and potentially self-sabotaging team at different stages of the project. Also, if your manager tells you that, you must look for another job asap. It won't end well, and you will never feel valued or heard. Even if you come up with a great idea to make the project, product or process better, it will be rapidly shut down or even better, the credit for it will be taken away from you and grabbed by that type of manager. They are the worst. Don't be that boss.

### And The (Not) UGLY Donkey

I didn't want to spend too much time on the bad because it is just a demonstration of how toxic a certain environment or people can be and should be avoided at all costs as it will always lead to failing to deliver what you set your mind on.

Finally, here was the story or tale told by one of my mentors and former boss that illustrates how to manage a team. Let's call it the Tale of the Horse and the Donkey.

Horses are magnificent. They have a wild spirit, but they can be trained and feel accomplished as part of a "team" or partnership. When jumping or scraping fields, humans and horses form a unique team in strength and intelligence. And a horse will go to drink water at the pond or lake when led to it.

However, the donkey is a different story. Most humans won't pay attention to this miniature horse, and donkeys can have a difficult temper or even be independent. They will show physical strength and carry a load, but they will do it their way, and if, for example, while climbing a mountain, they decide to stop or not proceed but go back, there is not much the rest of the caravan can do. Not the best team player. And there is a quote: "You can't lead the donkey to the water (to drink) if he doesn't want to go."

Now you see both. If it were employees, you would want to have only horses, magnificent, trainable and working as a team while following the lead.

As my old boss said, why discard the donkey? If it were an employee, it would be a pain to manage, constantly explore unnecessary paths on an agreed plan, and work more often solo than as a team. How to manage such difficult employees?

His answer took me by surprise. He said you want donkeys on your team because what they provide is not visible. They are the hardest workers going for long hours. They keep exploring new paths because they want to be certain the one they are on is the correct one (or easiest, smartest, or safest). By doing so, they don't question your leadership; they may just want to make sure they have your back and deliver their shares of work when it matters at the end of the journey.

So, as a leader, have both horses and donkeys. Remember, you can still train a donkey to go to the water, but you need to do it smartly, establish trust, leave some independence on time to drink the water and make sure that they feel as useful and beautiful as the horses.

Everyone contributes to a company, and great managers will recognize, mentor and coach everyone.

As a people manager, I have been fortunate to meet some of the best and smartest people in the industry, and some have worked for my teams more than once at various companies. The important part is when you find someone with a level of key expertise you cannot match, don't be afraid. They will contribute something to the team you can't.

The best story I like to tell was when I worked at Mastercard. The whole effort for the Masters of Code was coordinated between so many people, internally and externally, in different regions, that we had to keep track of who was in charge of what, where and when. We had our shares of "surprises," like when US citizens could not attend an event in Brazil, when half of our equipment didn't make it to Turkey, or when Google organized a competing event in the same city on the same day and developers were changing venues. But despite so many unplanned moments, and in retrospect, some were funny, I fondly remember the talent and willpower so many people put days in and days out, working long hours on weekends (it was not unusual to log 90+ hours in a week). Some of the regular crew came up with little hacks on the schedule or structure of events which made it the success it became. It is hard to pick the best "happy" moment which tops the list of 15 global events. All had a special feel. But two moments show what can happen when surrounded by a team where everyone has the back of each other.

Having the Mastercard engineering team involved in a tech event made sense, but it was hard to get some of the engineers to come to our events under the promise of little sleep, abuse at 3 AM by some developers when loaded with Red Bulls and sleep deprived or deconstruction of an API they worked on for months and that developers will willfully (or not) tell you how bad it is. For the first event, we got a couple of key engineers. Our promise was a trip to Australia, the chance to meet the people using their products and a lot of free swag (hoodies, t-shirts, etc.). We treated them like rock stars because we wanted the attendees at the event to be treated as VIPs or Rockstars, and our technical staff should be the same. We were under scrutiny by many pairs of eyes, and whatever we did would be seen. We did well because, at the following events, we had to create a waitlist for engineers to attend as they all heard about the "fun" and wanted to participate.

People are important, and if the team has chemistry either naturally, by trust, or by shared interest, you can deliver anything at any time and sometimes even more than you think.

Talking about that event in Brazil, we hit the biggest struggle of the series. We had the venue, the registration, the support of the local office, all government authorizations but just one. The Brazilian immigration administration had snailed down visa applications for American citizens in retribution for the newly imposed restrictions on Brazilians by the US immigration services. And we were in the middle of it. More than half of the core team was American. It didn't look good, especially when most were the tech staff. And this is when the team's brain power on how to make it happen came in. One of the techs suggested we could just get multiple monitors and create remote tech support stations with US staff helping from our office in St Louis despite the time difference. We inquired with the local team to see if it was OK and if the Wi-Fi could handle the traffic. Their answers were amazing. Mostly, they brought a full tech truck to the event to handle speed and volume. It may have been pricey, but my team never saw the bill, and the stations were a hit. But the best was to come. One of the participants had lost his two partners but still wanted to compete. Tom, one of our engineers from Ireland, was supposed to go on his break after working the graveyard shift (the 11 PM to 4 AM slot). Yet, it was almost 6 AM, and he still had to go back to the hotel to catch up on some sleep to be fresh for the team pitches at 2 PM. But he couldn't get the project from that guy out of his mind. He really liked the idea. As a staff member, he could not dedicate his time to a single team, but in that case, he was "off" and knew how to guide that developer to overcome his challenge. So here he was at 6 AM, helping and guiding this developer to finish his project. And that went on until noon. Tom decided he would just skip sleep because there was a lot of help needed on the ground and the remote stations could not handle some of these last-minute must-see-in-person bugs. So here he was, putting his well-being after the team's needs to ensure our participants had someone in flesh and blood to help. Beyond his dedication to the craft and developers, what was amazing is despite having not slept for almost 40 hours, when the event was wrapped, and we were finally sitting down to have a drink/dinner and debrief on how the event was, he came back at the restaurant after refreshing (i.e. we all needed warm showers and sleep after these mad days). He proclaimed he wanted to go and party in a nightclub. And he did. No sleep for 45 hours, and he is now known as Tom, the Beast of Dublin.

And this was not a fluke. I have been lucky enough to observe people managing their teams correctly. I have been part of such teams where you want to push yourself to deliver more because it feels like the right thing to do.

When I was CTO of Validity, we changed how biometrics was used. From a heavy material, centralized government type technology, we were trying to have some as light as possible from a material viewpoint and driven by mobility and decentralization. This is why we were founding members of the FIDO Alliance and provided some historical IPs to help accelerate that work. One of the biggest challenges in fingerprint biometrics is the capture of the image of the fingerprint for accuracy. Look at your finger and imagine hundreds of little dots delineating the most notable features on your finger. The problem we had was that the sensor comprised horizontal lines capturing as many as possible of these dots as the finger slid across the sensor. Swiping a finger on a mobile device was not easy; therefore, a simple touch was the way to go, especially when looking at an under glass solution. This was a requirement from one of the largest smartphone manufacturers in the world, and our deadline was the big event called Mobile World Congress in Barcelona. Everything had gone according to plan up to this point. We had the "sliding" solution ready to go, but they didn't like it (a last-minute change of heart as it was their original requirement, but they felt it would not be enough for that event. They wanted more). The team had convened, and we were disheartened because we couldn't find a solution fast enough to meet the deadline. Our materials engineering team, our software team, our components team, we were all baffled by what to do. Great team but stuck. All our solutions would either take a lot of money, increasing the cost of the sensor beyond an acceptable level for a consumer device or would take too long to implement because feasibility testing and research needed to be done, or it would require developing a brand new piece of software never done before by the team. Well, do you remember the donkey story? Here is a real-life donkey moment. One of the material team engineers was always quiet, not sharing much water cooler time with everyone. He had been sitting in the corner of the conference room for the past three days, not saying much. This was when he pulled the Hail Mary. In his view, we were looking at it in a way too complicated manner. We knew the horizontal lines scanner worked, and the software captured the needed image while distorting it due to the swipe. The thin "film" material used in the sensor was mostly a 10c component. In his mind, if we crossed two horizontal films together, we would create a "grid" of dots that could all capture the image needed. The software must avoid reading it as an elongated swipe but as a simple flat read. The match engine would need to be reduced on the surface; it needed to simply read the dots. The added cost in the material was just another 10c, some software modifications on an existing, proven match engine and voila. We all looked at him in disbelief. It was so devilishly simple, yet it was in front of us all this time, and we didn't see it. In his mind, we need to flip one thing around and voila. Design it simply.

And last, if you can identify a specific skill missing or the talent needed is not available for the specific timing of a release, and you still need to deliver, never be shy to ask for help outside your company. I have been lucky to meet external partners who provided more than what we bargained for and paid for. One lucky meet was Consult Hyperion. We hired them for some work at PayPal, and again at a few other companies I worked for. The reason? Because their experts are true experts at their craft but also not coming in as the top hired gun. Their approach is always more subtle, and they get the "team chemistry" fast. So instead of coming as a potentially disruptive force, they bring a proper invisible gravitas to any project. And for this, I have stayed in touch with a few members of this company over time. Dave Birch became a close professional acquaintance through these works. And having the capacity to talk to one of the top influencers and precise analysts of your industry regularly is priceless. Beyond his cheery knowledge of widespread expert domains relating to the field I am in (and publishing provoking books about it), Dave also provided countless opportunities to meet-and-greet other influencers and top executives in various industries. Besides having this lovely and unique British humor who can disarm an audience in two sentences, his sharp analytical mind cuts through a lot of the hype and smoke and mirrors of our industry and all these "latest entrepreneurs" darlings put forward. His sharp business acumen (and self-branding talent) has been something I have admired for a long time. So as you read along and reach the Chapter about "Cut the Hype", remember Dave. He can build up the hype and put it down in less than ten words. - Nick N. on the same team is a constant link that brings me back to asking for their help because he is always set up to deliver more than they agreed on. His approach is also to understand the dynamics of the projects, the circumstances surrounding the request for help and the team's composition. With a neutral approach, he can identify the best path to achieve delivery with the team already in place with as little friction as possible. And it comes with an invigorating chat with them at the end of projects. People do not realize how many projects in the finance, banking, mobile, identity and crypto worlds Consult Hyperion have delivered for others. When in pain to deliver your own project, they are the doula you call for help.

The capacity to find a great team and federate talents is one of the most critical aspects of being a delivery person. I cannot insist on how important it is. Honestly, I would not have delivered so many interesting products without the team I worked with, learned from, and led. The great reward is when you are with your team celebrating another product launch and the table behind you is asking the other person if they are using the apps your team launched a couple of months before. Seeing the success of the team in the wild and sharing this moment with them was one of these magic serendipity moments. I have worked and managed the same people at 3 or 4 different companies. I trust them, and I know right away how they will contribute to the team off the bat. However, I always look at teams I am asked to manage and try to find the nuggets, which never disappoints. Come with an open mind, and as a manager or leader, to deliver, your role is to organize the team to be optimal, trust their abilities, and then remove all obstacles in their way. Then you will deliver constantly.

### *Key Takeaway:*
Find your team and trust it.

# CHAPTER 3
# Dealing with Uncertainty or incomplete input

*"Should I stay or should I go" - The Clash*

The biggest challenge to deliver anything on time or within budget will always be the lack of clarity in either direction or requirement. Execution can be done against bad input, but it won't lead to the expected result.

We initially discussed having executive support or sponsorship to help remove obstacles. This is also critical to clarify what is expected to be delivered. Sometimes, these requirements will be generic or super detailed to the point of having a list of Key Performance Indicators (KPI) and Objectives and Key Results or (OKR) to be achieved. Whatever it is, this will be your starting plan for how to deliver and what. This is your peripheral circle of operation.

Beyond executive endorsement, if you are working with a visionary or innovator, they may share the big stroke of what's in their mind, but it is a different story most of the time when it comes to the details of the execution. This is when having a close relationship with that person and getting to know the basic thinking process is of great help.

So you got a general idea of what needs to be done. Now, the other element that helps with uncertainties lies in two parts. Internally, having a good understanding (or at least someone who can confirm one way or the other) of the overall corporate strategic roadmap will be supported by the project as it makes it easier to check if you are aiming in the right general direction. Nothing is worse than spending months on a project and being told it was not strategically endorsed. Also, be careful; strategies often change along with executive sponsors. Regularly check these to ensure a 180-degree turn was not done with little fanfare. Sometimes changes are subtle, and the impact is difficult to discern. Having regular touchpoints with other teams, peers and managers always help. If you never ask the questions, you will never get the answer. I learned from a couple of bosses that they are never good at reading minds and will let you know. The second aspect is to have a great (or at least good) understanding of the ecosystem in which the project needs to be delivered. Is it the intersection of multiple ecosystems? Then look into your network and ask people what they think of this or that. It is your responsibility to keep your knowledge current, and there is no better way than watching videos, reading articles, and attending conferences. If you are lucky, you can do these on your company dimes, but don't be shy about doing it on your own time. I had so many informal chats with people I admired during evenings or weekends. Keep your work-life balance in check but don't get stuck on a routine either.

The best way to explain is to use examples of how uncertainties turned into clear marching orders.

The team was small when I was working on the mobile contactless payment solution for PayPal Mobile apps. We did a lot of "skunk" projects not endorsed by management but tolerated for their potential impact. I had gotten a few diverging opinions or comments from executives. The biggest one was from the eBay CEO, the owner of PayPal, who declared in an interview that NFC was "Not For Commerce." Now imagine you are in my shoes and holding a phone prototype with contactless payment, including NFC, in your hand. I was asked to deliver as many innovations as possible for proximity payment (which generated a lot of patents), and now, the messaging was mixed. When I asked my boss if we should continue down the path we were engaged in, his answer was to keep pressing but not waste resources or ask for more budget. It was an indicator that I would be on my own if things went south. I was working on the original idea for PayPal to beam money from one device to another without wires. So it couldn't be that far off, right?

At least the advantage gained was that the mobile app was a success, and the transaction and usage velocity were going up exponentially. Mobile was on fire, and I was part of the original team. So it was easier when reaching out across different teams to get goodwill to volunteer to help on some specific part. I was operating in murky waters but determined to show the executive team the completed device with all the bells and whistles embedded in it.

And part of the mobile app launch was to attend many events as a speaker or just a VIP guest. It was a great time, and when I landed in Singapore to visit our main Asia hub, I was amazed to have the head of the APAC (Asia Pacific) region ask me to come and see him in his office. He would be excited about what I talked about and to see all the great work achieved over the past six months. Mario S. was originally from Central Europe but could have been a citizen of the world for as far as I could tell. His impeccable manner and English projected executive presence, and being in his office was intimidating, but I was part of the mobile team, so it was all good.

Well, it didn't go as planned. The original meeting was for fifteen minutes. It was almost one and half hours. His first words were not what I expected the discussion to be. He bluntly said, "So, I am hearing you are using two of my engineering resources without any approval? Can you explain to me why and what you are doing?" And so I was now in the hot seat, uncomfortable and probably red-faced. If the human body could morph into the chair I was sitting in, it would have been the perfect time for it. At the end of the meeting and all my circumvoluted explanations, Mario had a reaction that, to this day, earned him my unlimited admiration. Paraphrasing his words, he said, "Well, I get it. It's quite important. You can have five of my software engineers but not one more. Is this clear?" And from there, the project took an accelerated path. When Mario offered to join at Mastercard to drive the Developers' effort, I didn't care about the offer. I was just super excited to work for him directly this time. With patience, he listened to my explanations, analyzed the overall risks and benefits, figured his team could help and lifted a lot of timing and resource uncertainties in a few minutes.

The other challenge can be linked to choices made for a project before you join a team and you have to make it work or change it. It is even more relevant when linked to technology. When we were working on Zelle, it was a new product resulting from the merger of three companies to deliver the first real-time payment network in North America. The main company, Early Warnings, is a joint venture between seven large banks in the US. It is easy to figure out that whatever technology decisions we were making were always questioned by the CTO of each bank. Since we were building something that had never been done before (while trying to integrate the underlying legacy technology of three very different companies which were merged while supporting the technology stacks from each bank), a lot of unknowns and uncertainties were in the balance.

Combining the release of a new solution built on top of three disparate technology approaches while not breaking legacy solutions was not a small feat. Being a real-time financial service implied a lot of scrutiny from the Governance and Regulatory Compliance team AND infosec team. The risk was major and the first answer from these teams for whatever you ask is always a firm "NO." As for a few other projects in my career, the first order of business was to talk to the legal and compliance folks and try to understand the margin of movement. In parallel, the discussion with the infosec team was also ongoing, but I had more confidence in that aspect as my background was similar, and that team was especially top-notch and willing to help to make it work. After a few weeks of back and forth and helped by a new Risk Management executive also putting in place some parallel safeguards, we were on.

The biggest challenge came around the Data Policy to follow. Anything using consumer data to trigger a real-time, irreversible cash transaction was not to be taken lightly. The debate was to figure out if we were going to use cloud-based components or continue with physical data centers (with armed guards, yep).

The opinions were divergent internally and in our interactions with some banks that didn't want to change their infrastructure. If you think managing one project budget and deployment lifecycle is hard, try eight. We even had a bank asking us to create a new way to do API calls by turning things upside down. As you can guess, the team pushed back, explaining why it was a very bad idea to start with and an unsustainable long-term approach as it will never comply with standards and open architecture. We settled on a shared directory system partially relying on some of the risk systems from the banks themselves but, more importantly, on the AI and Machine Learning technologies developed internally at Early Warnings. The best (and most surprising) ally came in as the head of product for a smaller bank, who was a former CTO and got to always reframe what we wanted to do in the banking lingo for the other members of Zelle. It was amazing to see potential tricky moments turning into decisive actions, removing any doubt or uncertainties at the executive sponsoring level in the process. On the technology front, it was always managing a lot of question marks, but we launched Zelle in 2017, and it has been growing since then.

In your plan to deliver the best product possible, include some (small) buffer for delays. These can be legal, financial, technological or whatever life can throw at a team. Sometimes it can be an unplanned event. I had a shingles crisis during the Mastercard Masters of Code. While recovering, I forced myself to attend the event in Singapore. It was the most miserable and painful time for me and being an unnecessary burden for the rest of the team. They knew what needed to be done, everyone was ready to take over some of my role, and I should have just stayed home and rested. I wanted to make sure everything would go according to plan. Maybe because we were early in the series, I had some insecurities about how I planned the series, but I should have acknowledged after three events that there were potentially two other delivery people on the team that could fill in on short notice.

But sometimes, uncertainty comes from a dark moment no one wants to deal with. Sadly, in one of my early careers, the company I worked for lost their CTO, who accidentally drowned while white rafting. This is the ultimate unplanned event. Even with all the best continuity plans, it creates an additional element. Emotional. Again, accept it. Genuine empathy to others is not a weakness and will make you a better delivery person by anticipating these or at least knowing the right thing to do or say.

It was publicly shared, so I can also talk about it here. When in the final phase of relaunching a new developer experience for Envestnet, Jud Bergman, the President of the company, was excited, and we had planned to have the pre-unveiling of it to him before Thanksgiving freeze. In one of the saddest moments I witnessed in my professional career, Jud and his wife Mary were killed by a drunk driver in October. While Jud was not involved directly in my project, he was the key executive sponsor. His vision was guiding us to where we wanted to be. The Business Continuity Plan was activated, and business was back on track, but to this day, I always wonder what would have been Jud's reaction to our delivery. I always kinda hope that we delivered on his vision, and a bit of his influence was (and is still) seen in the project.

### Key Takeaway:
It's going to be messy. Just accept it.

# CHAPTER 4
# Dealing with negativity.

*"Fight for your right" - Beastie Boys*

With success comes envy. And with envy comes negativity. It's easy to let that get to you, but here are a few tricks on how to avoid letting toxicity make your job miserable. Without passion in what you do, you will fail and negativity has a talent to drain the passion out of most of us.

In the early days of my career, more than once, people took credit for my work and didn't even acknowledge my contribution. The most likely culprit could be the visionary you helped transform their idea into a reality. Here, it can be explained by either visionaries having a big ego and narcissistic tendencies or because the empathy and social skills are not fully present. Awareness of the contributions of each team member would, most of the time, fold onto you. Your job will be to whisper to whoever can recognize and award these people the little extra bonus or some extra paid time off to recover, etc.

But visionaries or innovators with limited awareness are not new. It is even expected in Silicon Valley that all "geniuses" are not known for their people skills.

However, another aspect of negativity will come not from the people you are trying to transform the vision of but from people supposed to either support your project or at least be encouraged by it at the corporate level.

The types of negativity you will face are no more than four.

- *The Skeptical:* Most of the time, this will be at the executive sponsoring or program management level. Recently, my team had been delivering back-to-back products to support developers. This gave us a lot of executive-level visibility but scrutiny comes with visibility. One of the operational executives looked at some data analytics. My team also had quite an advanced system to track all activities on these products (with consent from the users). Without consulting us on his numbers, he had discussions with the finance team and told them we were a cost center going out of control. In his mind, this was the proper way to look at it. How much budget is assigned to the project, and how much revenue comes out of it. Quite a normal operational instinct, especially when your background is in accounting.

What was missing from the data points being used was what I like to call the "soft credit." A developer program's main focus is not to just generate revenues only. With the proper platform in place, it helps onboard users faster, allows them to find information in a self-serve fashion, reduces the cost of supporting staff and, more importantly, allows them to distribute, publish and improve products at a faster pace. Large customers will sometimes come through the same path to cut short the required integration time or because they want to scale at their own speed. At one of the large companies I worked for, a bank renewed a contract for multi-years and multi-billions of dollars because one of their top ten reasons to stay with us was we were highly innovative, as demonstrated by the developer program, which was their #3 reason out of ten. I'll take this any day.

So, the best way to deal with a skeptic is to listen and see where they are coming from. The best bet is always from a financial viewpoint or lack of understanding of the market/customers. I have learnt over time that if you take the time to present factual numbers and educational input, it will get you far in converting the skeptical into a believer or neutral party. If you convert a skeptic to become one of your champions, they will always go at bat for you. They can be the fiercest defender of your project. Mario (chapter 3) was skeptical when we first met. And yet, he became a strong advocate and supporter of my projects.

- *Envious/Jealous:* This is the worst for me. While the other negative aspects are rarely triggered voluntarily or can be perceived to improve, the envious will try to take full ownership of your project, even if it means pushing you out of the door of the project or the company. It happened in my career more than once. There is not much you can do besides pushing forward or moving out of the project/company. The behaviors of a person with such feelings will not be professional. Even if you keep repeating to yourself we are all professional, we are all on the same team, we are all working towards the benefit of the company, guess what. Some people are not born to be team players or good human beings. The interference can go from light derailment (moving a critical resource to another project) to backstabbing your project (if not you) with other leaders, demeaning your reputation to plain and simple sabotage "Oops, we decided to change the vendor for the underlying platform at the corporate level. We didn't tell you? We should get the new one up and running in a year or so." It is not rare to see an escalation in the wrenches being thrown your way. With experience comes the capacity to see little early signs of it but even to this day,

I am still sometimes surprised by the behaviors of people I thought were allies. As a politician (Reagan) said a long time ago. "Trust but verify."

I was leading a large project when we were perceived as the most innovative team with the specific audience we were targeting. Unfortunately, we were part of the core product side of the corporation, not the innovation laboratory team dedicated to generating what we were doing.

An easy solution would have been to integrate us into this other line of business, but one of the top guys for that research laboratory convinced management to let the whole team go. We were in full ramp-up when our wings got clipped badly. It created some reputational damage with some of our users, but the envious didn't care. He had removed a threat to the standing of his team. Only a few months later, he was out of the company too. C'est la vie as they said in America.

- *Clueless:* How do you deal with someone parachuted into one of your projects but is out of their depth? It happened most of the time in large companies with an international flavor. I remember one time working under contract for a large semiconductor company. One of the executive champions in the US bypassed the recommendation from his peer in Asia and hired one of his friends as an employee. That person was assigned as "co-lead" of the project but being an employee and me being a contractor, you can see where the problem lay. And to add injury to complication, he was not well aware of what we were doing, not hardworking to learn. He thought I was his staff or secretary or slides designer against the terms of my contract and therefore would have to be renegotiated. We delivered the product, got it certified, and it was well received by the market. A large marketing campaign was launched, and while I was doing most of the customers' visits, he was going to talk at conferences despite misstating some basic elements of the products (wrong certifications, wrong country coverage, etc.). This was frustrating for me as we were already two years into the project when he was dropped in it. It could have been catastrophic, but we pulled it off. While his attitude didn't change, he still saw me as his luxury secretary vs. a proper program lead. After I mentioned it to a manager from Asia I was closed too, the company hired me as an employee to solve the problem. The biggest issue was they also had me report to that person. We went from being semi-equals to confirming the hierarchical assumption he made (and that was never communicated to anyone). When the next project started and I saw how it would be, I left. The project failed miserably as it didn't even take off from the research lab slacks.

In another company, we were small and trying to stretch dollars as far as we could. Our investors were adding funds, not as VCs but as bridge loans until the next round. It was a constant chasing of money to keep going. One time, an investor put a lot of conditions on the next loan. This included slashing salaries for everyone. I was not paid a lot at the time, but it hurt to see this 10% going away, then another 10%. And our stock options didn't seem to add up to anything. The next recommendation from the investors was insane. He parachuted an interim CFO to monitor whether we were spending his money wisely. The team was already demoralized from the pay cut and now, we had a money chaperon. It didn't go as planned. While he kept telling us we were not working long hours enough (he was not there on weekends when we were in) or that we could find better ways to stretch our already thin infrastructure (secure servers are expensive and need to be updated often!). The morale and atmosphere in the office were not good, and we couldn't figure out what values he was bringing. He didn't even know how to proceed with payroll. We had to get our former office administrative person, who was let go when he joined, to come back and proceed with payrolls under contract. This was not the change we needed or how to motivate the crew. But sometimes, there is justice. The "genius" interim-CFO was caught surfing porn on his computer and sending inappropriate messages to friends via corporate emails. The reason he was caught is that in his willingness not to release more money for our infrastructure, we had to add a lot of monitoring tools to it to help with capacity planning and peak monitoring. It applied to our complete infrastructure including mail servers. He also wanted to check on our "emails" to see if we were doing real work while at work. The sniffer he had asked us to install worked for everyone. What more can I say? It was an example of dealing with cluelessness to the top.

- *Changes:* The last one is something you will never get control of. I debated classifying as a negative or an uncertainty belonging in Chapter 3. It can be a change in the project, team, resources, budget, strategy, company direction or a company going belly-up (I had my share of start-downs). The biggest change you can and have to anticipate is an executive sponsor leaving the company. While it may have provided some air coverage for a while, you can never assume that a sponsorship is permanent. This is why it is important to talk to many people about the project's progress and gain insights into corporate strategy and directions. It will greatly help anticipate a change and prevent its potential impact on your project. In the wider scope of negative events, this is the lesser evil because it is not yours to own if it is at the corporate level. However, everything within your project delivery roadmap is yours to own as stated earlier in this book, always run a tight ship but be able to respond to a change rapidly to minimize impact.

When Zelle was being built, many questions were still floating as we were moving forward with the technology. Many aspects were considered, from the underlying platform to the risk engine to the mobile apps experience. The original marching order stated we had to get a stand-alone Zelle app for onboarding as many users onto the network as possible. As we were pressing forward with the mobile development, we were pleased with the overall UX, the design and the flow guiding the user step by step. As we were getting closer to launch, we reviewed with all the owners' banks. While they all seemed pleased with it, we also learnt that a couple of the bank executives would move on to other projects. This was a bummer as one of them was technically inclined and always a good sounding board. This was a big change and a sponsor gap we thought little about. And then, reality hit hard. As we moved along the authentication flow, one of the new tech execs noticed something in the stand-alone apps. He asked why our authentication seemed to have two steps reversed compared to their integration into the bank apps and probably all the other banks. We were so nose down we didn't isolate the stand-alone apps with direct access to our platform, with the ones of the partners having to integrate indirectly with our platform and authentication system. As you can guess, we were left scrambling and re-ordering our flows as it was out of the question to have a dozen banks integrated to re-integrate and change the flows of each one of their respective apps. It would not have flown far away. Because the team had a lot of great talent and our authentication experts with the acquisition of Authentify, we were able to turn around and modify our own before it was too late. I will spare you the long discussion with the risk team, who were not happy to see all the security steps they had in place being partially switched when they were also close to deliver. This was a change we could have anticipated with more communication and more regular touchpoints with our "banks" sponsors. Even if we had meetings often, we also had some flexibility in our implementation. This was a short sight because the project was kept in a silo instead of being in touch more often with the other partners' tech team. We were so paranoid about having the project leaking before it was ready that we had too many safeguards partially blinding us. That one was on me.

There are also changes you must do to protect your team. If someone is toxic or is not motivated to deliver at the speed of the rest of the team, you must make the change. Or to leave to make room for a change.

At a start-up in my early Silicon Valley days, we were working on securing any devices with the capability to accept payments. Our embedded software had a small footprint, and was easily deployed, but our team was small. The technology was a spin-off from a larger project at another company. While we were prospecting for partners and customers, we were hoping to get a big name to help boost our revenues and chance of survival. We were in a meeting with a large financial network when our CEO did a deep dive into the technology. We had already asked for a Mutual NDA (Non-Disclosure Agreement or Confidentiality Agreement), and it was agreed upon. However, when we arrived, he asked for them to sign our specific NDA because we were going deeper than originally planned. The other company's team explained that any legal change to the NDA would require a review that could take a few weeks, so we had to either proceed to the level of what we could disclose or postpone the meeting. How do you go from a potential partnership into being almost kicked out by security to the curb? Threaten a large company with lawsuits and patent infringement when the conversation has not even started. The team was disheartened, and after that meeting, I knew it was impossible to change the CEO, so it was time for me to go.

Just remember that in Silicon Valley, few CEOs would be deemed as psychology "unstable." I had my fair share when I was running my consulting firm. Some are great visionaries and motivated tech people with little social interaction capabilities. If they are smart, they usually get it. They need someone to take over the 'day-to-day" relational part while focusing on what they like to create.

A CEO I know went into a major disagreement with potential investors. Instead of trying to discuss potential terms or changes to the term sheets, he banged his fists on the table and told them it was his way or no way. As you can guess, both VCs stood up and left the conference room. One of them dared to comment that his attitude would probably blackball him among the VC community. He went so mad he threw his Montblanc 149 at them. Apparently, the pen pierced the drywall right between the two VCs above the elevator button. Their eyes were wide opened, and after a couple of (deserved) swear words, they disappeared into the elevator. Enough to say that the start-up never completed its financial round. Even with a CEO change, the name was banned from most VC firms in Silicon Valley. When the toxicity comes from the top, the only change is for you to leave. The importance of Emotional and Intelligence Quotients cannot be underplayed in today's world.

Meanwhile, the positive is that I have also worked with terrific CEOs from whom I have learnt a lot by observing their decision process, how they treated people, and how they behaved when no one was watching. If you can find the gem, the first thing that will be obvious is the level of respect. The second important part will be the natural establishment of trust. Lastly, with respect and trust, you will have the maneuvering capability needed to make the changes you need or even the ones you want.

There will also always be a certain level of negativity coming from a failed project. We have all worked on a project that may be canned or a product release that didn't live up to the hype. You must own these failures and learn from them. Doing a Root Cause Analysis is always good, but it is always too late by that time. There is no more sickening feeling than to call for a roll-back at 6 AM for a web-based product roll-out, especially after the team has been up for eighteen straight hours. That call will have to be made at a point, and you will have to own the explanation. A few times in my career, I had the unpleasant duty of explaining to my boss (the visionary) why her/his/their vision ended up in the gutter. If you can demonstrate that everything was done to achieve success, yet unforeseen events made it impossible to accomplish the delivery, any reasonable person will ask if it can be fixed and a new delivery date found or maybe if it can be recycled into another project. This happens. However, keep in mind that during this time of a failed launch, you will become a target of people who are more than happy to shine a negative light on your accomplishments or methodologies. How do you deal with this? Again, get strong executive support and communicate, gain respect for the decision you made and make sure you have others who may be willing to step in and downplay the stumble. It will be a difficult time for you and one when you may lose faith in your abilities but always try to remember all the times you were successful. One failed launch does not reflect on your overall professional journey as badly as you may think at that moment. If you think that way, it will freeze you in not making the right decision even if it negatively impacts you in the short term.

Sometimes you cannot do anything about it, or the results are not there despite implementing some changes. Things do not always work, but even more in an unsupportive environment. Also, learn how to pull the cover back to you. Communicate with executives, and make sure your achievements are noticed. Being a foreigner, it was a hard thing for me to do. Mostly, I was paid to do a job; if I was doing what was expected, then people would notice. Nope. In Corporate America and especially in Silicon Valley, you have to be your own champion. Don't be shy to remind people what you have contributed or that you were a key person in a project. Otherwise, someone else will take the spot. As we say, let's give back to Caesar what belongs to Caesar.

### Key Takeaway:
Don't let toxicity and negativity bug you down.

# CHAPTER 5
# What's the secret?

*"Over the Rainbow"- Israel Ka'ano'i
Kamakawiwo'ole*

The moment I realized I was a good Delivery Man was not in a single instance. It was over a longer period. I had instinctively been doing the right things but never in a formalized or structured fashion. There were patterns over various jobs and companies, but my mindset was not on finding the commonality. The start of the epiphany was at PayPal. I started observing how people were doing things in areas that were not my expertise. Then, a few companies later, I attended a happy hour event for venture firms and entrepreneurs in Silicon Valley in 2017. I was out of Mastercard, still riding the success wave of the Masters of Code, AND now being the Chief Technologist at Zelle, people wanted to chat with me (a great feeling but not to be abused). While discussing with my brother-in-law Noah (a crypto entrepreneur) and another person, I was trying to explain what I was doing. So I described the projects I worked on. While in my mind, it was more Forrest Gump moments being there at the right place and the right time, that person exclaimed, "Wow, you **delivered** some great things".

Finally, the dots connected. All my previous roles, my previous behaviors and role ethics came into a single light. Even my earlier concept of Innovators vs Integrators made a lot of sense. The certitude of knowing what my superpower was, was exhilarating. While I always considered my career bouncing from a company to the next thing I wanted to work on, now it had a single cohesive thread. Afterall, all these people complimenting my career on LinkedIn or asking for advice may have seen something I was not. Now, all I needed was to structure it a bit better to understand myself and also why I was doing things in a certain way, and it seemed to be working repetitively.

I started analyzing all the past projects and the current one I was working on, trying to find some magic repetitive patterns which could be the silver bullet. From an operational point of view, there was not much. Projects had been different, team composition was not the same and the executive sponsorship and budget had also been all across the board. So I had to look further into the biggest variable. Human behavior and psychology. Where was the talent in getting things out of the doors? And it was not the quest for perfection. And then, another past practice came to light in a very different context. Now, having the capacity to think about my career with a wider lens was an amazing eye-opener. The dots were not only connecting. They were leading to the next ones. I had found myself professionally.

Therefore it is not really a secret but a way to approach it with a specific mindset. Great program managers instinctively follow what will be described below. I met a few in my career, and when mapping their behavior against a specific thinking methodology, it became clear all of them had overlapping behavior allowing them to bring a project/program to its term, which is the delivery.

Linear thinking is a good start but not the secret. Being a good planner can help but is not it.

Being reactive to obstacles along the road is great to stay on course but not the big secret.

The important part is to visualize the finished product (it being an app, an electronic board, a phone, a car, a rocket) and share that interpretation with the visionary person. When the underlying understanding of "what it is" is to be delivered, then deconstruct what you see as the finished product into smaller pieces. Order then in such a way you can overlap deliveries, add uncertainties in timing and dependencies on other factors, get a potential timing agreement with everyone who needs to be involved (everyone hates the "Exec A promised the shareholders, the board, customer B, whatever that the new product will be out in [insert unreasonable date here]" and check it. Being a great delivery person means being good at program/project management, requirements crafting and tracking, team motivating, and keeping your nerves in check.

The best way to achieve the above is to follow a methodology called SSOCCADD. When at PayPal, Upendra Mardikar and I took this military training concept and applied it to technology security to mobile devices (you can still find articles in a few places, but it was back in 2012).

This methodology applies to a lot of processes in the business world. If you train your brain to think that way, it will become automated, and you won't even know you are doing it. This is the reason it took me some time to figure it out. I was doing it not on cruise control but without thinking about it. Consistent behaviors and talking to other great delivery people helped me connect the dots.

When dealing with an unknown or unfamiliar environment, find new bearings rapidly. When under the pressure of combat threats, achieving this in a critical, timely fashion is even more important. This military principle is applied to the corporate world and can be simply summarized by the acronym SSOCCADD explained below.

Anyone in the military around the world receives this type of instruction. Memory tricks are associated with repetitive training, so the mind can focus only on critical decision making while the prepared repetitive habits kick in to take care of the "predictable." Sometimes, it may save your life. In the corporate world, it allows you to instantly view the "big picture" while rapidly understanding the details of it.

While discussing among colleagues, a consensus emerged showing that the same principles could be applied to a company's business, technology or security process. One area of focus in the technology world is always covered by one of the three areas above. Hence, the primary motivation is to introduce the SSOCCADD principle to a large audience and, secondly, to provide concrete applicability of it in the most exciting fields for various industries.

But before going into the flow of the process, let's first define the meanings of SSOCCAD.

## The principle

SSOCCADD is a situational principle developed to understand in a structured way the uncertainty of a situation in which a person or a thing is projected.

The meaning of each letter is as follows:

- *Situate:* when working in a new environment, situate where you are at a macro level.
- *Secure:* not knowing if the environment is friendly or foe, you must secure your position rapidly.
- *Observe:* when the environment is secure, it is necessary to observe the immediate surrounding in more detail to refine the understanding of the situation
- *Control:* as soon as the environment is understood, it is now critical to control what is deemed important within the environment.
- *Communicate:* now that a position has been established with clear perimeters and critical elements, this position should be communicated clearly to your immediate

surrounding team and other teams supporting your position.

- *Analyze:* gathering and communicating information are important. However, when in new surroundings, the proximity to such an environment can provide firsthand information that should be analyzed on the spot, making you the expert. Teams supporting you may also start their own analysis of the situation.
- *Distribute:* whatever or whoever is conducting the analysis, the results should be distributed among the immediate environment and supporting teams.
- *Decide:* having observed, analyzed and distributed information among multiple key stakeholders allows reaching out rapid decision making that can affect the perception of the environment and/or generate an adjustment to the position.

When looking at the SSOCCADD methodology or principle applied to the world of product delivery, it is easy to look at it from the main streams of work usually found in a technology project. Mostly, the three main streams are as below:

- the business stream covers the project definition, requirements and product scope
- the technology stream, which is the development part of the product by engineers or mechanics based on the product requirements
- the security stream or in the world of technology, the infosec (information security) and/or pentest (penetration testing) teams along the QA (Quality Assurance) / testing team.

Let's look at these in more granular detail (figure 1)

- *Business Flow:*

Applying the principle of SSOCCADD to the business decision-making process results in a reverse read of it, especially for C-Level people. The decision is made (sometimes using partial SSOCCAD flow) and then distributed to various teams. Each team will then analyze the impact of the decision, communicate to the lower echelon and make sure that they retain control of how it is implemented. After observing how it evolves, the team leader will be able to secure the achievements of how well the decision was understood and situate the team into the next round of business endeavors (most of the time acquiring more budget, being promoted, being granted the management of a key project, etc.)

- *Technology Flow:*

For the technologist or the engineer, the SSOCCADD flow is, however, linear. For example, facing the development of a new product or the coding of a new program, an engineer will try to situate the scope of work, secure the best path to achieve the main goals, observe others to see if better paths exist, control the development at the tech level or tools level, communicate the progress to other teams/engineers potentially involved, analyze the first or subsequent rounds of development and distribute the final "product" for testing, quality review and integration. A decision is made to improve, freeze or maintain the project/product.

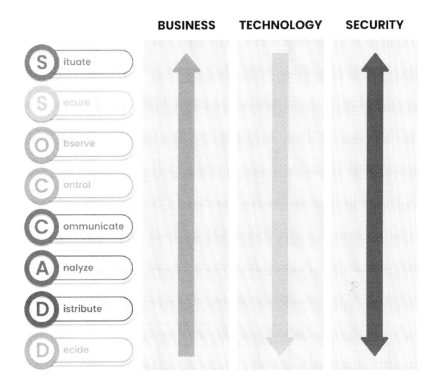

*Figure 1: The three linear flows: Business, Technology, Engineering*

- *AppsSec, Infosec, Security Flow:*

SSOCCADD becomes interesting during the application to the technology security world. As it was developed for the security of military personnel, it is well suited to be applied to the world of technology security but more especially to the new and unknown world of mobile and distributed security.

While reviewing the flows for business and technology, it becomes apparent that in matters related to security, the flow is dual, constantly oscillating between one point of the process to the others.

To understand technology security and to develop a proper strategy, one must understand both the "business" ecosystem in which the security is used as well as the "technology" parts used to support the ecosystem.

SSOCCADD is not a linear process but a circular one (Figure 2) that allows various functions inside a corporation to cross reference on well-understood points (let's say Control) and focus on solving issues that may arise. Business process meets Technology implementation meets Security requirements.

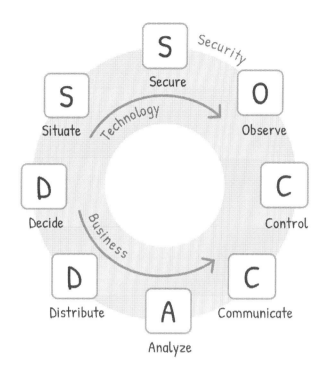

*Figure 2: The Delivery Patterns of the SSOCCADD flow.*

Now that you have learnt about the "secret," can you apply it to projects you worked on? The ones who succeeded but also the ones who didn't? For these, what was the critical point of breakage into the SSOCCADD process? The answer is in the middle. The most common point of failure is not paying enough attention to the two "C"s. Control is important, but more importantly, Communication constantly and clearly is critical.

Let's try to integrate some key questions into the SSOCCADD methodology to bring into context how it can be applied to elements that appear in this book or real-life projects.

- *Situate:* What is the strategic scope/goal of the project?
- *Secure:* Is there an executive sponsor? Is the project funded?
- *Observe:* What are the true drivers and motivations behind this project?
- *Control:* Are the scope, schedule, and cost/resources well-defined and understood for project implementation?
- *Communicate:* Is there transparent communication between the project manager, business analyst and, more importantly, the stakeholders or executive sponsor?
- *Analyze:* What are the impacts, and risks of changes to the project, aka scope creep?

- *Distribute:* When changes are inevitable, are the right teams aware and involved in implementing new tasks?
- *Decide:* When changes are happening around scope, schedule, or cost, can you make the call with a clear vision of impacts?

And as mentioned earlier in this chapter, the SSOCCADD doesn't have to be linear. You should re-visit these steps in whatever order you deem is important for a particular aspect/moment of the project which needs an answer. You can even define your own questions against SSOCCADD to achieve your delivery style. Nothing is ever set in stone.

It will make you enjoy the rewards at multiple levels. Having a structured approach to delivering your next big thing and to see a product you work on used in the field.

There was nothing more rewarding than seeing college students paying a tab via the PayPal mobile apps back in the days when few options were out there. Or when the cleaning crew at my home asked me for the first time if I could pay them using Zelle and if I knew how it worked. Or when the first Samsung phone was unveiled in Barcelona using our sensor, beating Apple by a few months.

*Key Takeaway:* Common sense is the best approach to deliver but having a methodology to help check project health is the ace in the sleeve.

# CHAPTER 6
# Cut the hype.

*"Dizzy" - Tommy Roe*

This is the scary part of entering into my mind. A good delivery person will cut the hype and go to the core of what needs to be done.

It's very easy when you listen to top innovators or visionaries to feel inspired and taken into the enthusiasm of the moment. This happens to everyone. You are in a new job, you have a charismatic leader, and you have been put in charge of a big important project for the company. The temptation to go all in and make it even bigger is easily a normal human nature reaction. However, words of caution cannot be strong enough when you want to be the best delivery person.

How do you plan for that big visible project while staying grounded in the reality of the business? In my career, I have been so many times in that spot it may seem easy to do with time, but you will still have stomach knocks when your CEO announces to Wall Street in an earning call that your project is one of the top three priorities for the future of this company. Talking about being put on notice that success is mandatory, failure is not an option.

A great delivery person can cut through the smoke, noise and conflicting marching orders to focus purely on the product and the process of getting there.

We explained earlier the SSOCCADD methodology, now try to take it to a more granular level. Under which step, what else needs to happen?

Coming from the world of technology, we have to look at the obvious scoping vs planning dynamics, which constantly require balance while sometimes overlapping functions or unclear role definitions will introduce more noises. However, these are human resources problems. Everyone follows a different emotional discipline, but there is another tool the delivery person will constantly use to help "organize" priorities and human roles to achieve success. I cannot avoid linking this methodology to the System/Solution/Software Development Life Cycle (SDLC). It would be great if everyone could agree on the meaning of that "S" and how many steps/stages/phases are included in a proper SDLC. Do we go with 4, 5, 6, 7 more?

It is a personal choice, but I like to use the complete approach of looking at the planning, requirement analysis, design, building, testing, deploying and maintaining. In there, you also must throw documentation. It doesn't have a specific place in the process because it should be done at any stage. It is a continuing effort, not a punctual one in the cycle.

In the SSOCCADD method, engineering will go the opposite flow of the business in planning. So depending on which side of the aisle you are part of or the nature of the product/service to be delivered, you may have to map the SDLC from one end or the other.

Now, let's see if we can get some concrete application to it.

- *Analysis:*
  Never start planning for anything if there is no good reason to do so. Is it an executive mandate? A product strategic need? A market demand? A key customer request? A sales team asks? Whatever the reason, there should be a clear market analysis of the perceived and potential benefits of doing such a project, especially if the cost to do so will be significant. Be careful not to fall into the trap of being the one to create this analysis. As a delivery person, you should not be focusing on that. We all know salespeople begging for a special feature to be added because a large customer needs it. Still, when you ask how much potential revenue this customer will generate, a cricket sound replaces the enthusiasm because no one did the analysis. If the requestor has not done a minimum of homework, then politely ask for more information. Even if it is the CEO, especially in publicly traded companies with a strategic

team thinking about the next move. That move must have been documented somewhere. The answer "because I told you so" is never acceptable (see bad boss section). Be assertive and polite in asking for justification to your time being committed for a significant period. This will cover the first two "S" from SSOCCADD in the way you should now understand the "why's." It also covers part of the "A" as this primary analysis will help you start. However, never stop analyzing, where your project stands along the way. It is important to narrow it down to one and only one goal. Increased revenues, customer retention, customer acquisition, cost reduction, operational efficiency, etc. Pick one and focus on it as the main driver of all decisions. Congratulations, you have the business requirement done.

- *Design*
  Design is one of the most underrated portions of planning and yet one of the most critical. Engineers should never be left in charge of design. Ever.

As you are working on designing the product, visualize all the flows and potential branches. At PayPal, we used to have mobile screen workflows printed on giant pieces of paper. From there, we added strings linking to engineering steps and requirements. So whenever someone was "moving" the screens to improve the flow or experience, we could see all the strings illustrating the technology dependencies moving along. When a LOT of strings were moving, you can be sure we were asking the question "Are you sure we/you want to do this?" more than once. It was a very visual way to measure impact.

Fortunately, the time for strings and papers is over (almost), and today, you can partner with designers to help with tools like Figma, etc. The discussion between product managers and designers is the most important because it will lead to the overall roadmap and final planning. I have been fortunate to work with great design leads who contributed significantly to making products better while adding little overhead. When working in payment or developer technology, your end-user is the person you need to focus on most. At Envestnet, we had a poll of design talents, probably some of the best in the industry. Besides designing the user experience flow and sequencing, they also had experience in some obscure areas or were former developers, which is extremely useful when designing developer experience. It cut the discussion cycle by a lot.

I was fortunate to have the same experience with a designer lead who worked at PayPal, Mastercard and who has turned poor design into award-winning solutions.

A pre-documentation review must also support the design cycle(s). What will be needed to explain the product. The design will be the beginning of the documentation team engagements you will need to have along the project.

As you have worked on designing a product's visuals, flows and sequences, you must ensure you can explain it in the simplest way. Why?

Because you now have the "what" very well understood, and it also covers many steps of the SSOCCADD method. After the design phase, you have started on the path of Observing, Deciding, Controlling, which leads to Distributing and Communicating.

Congratulations, you have the product requirements done.

- *Building*

  This is the most dreaded phase of any project for any technical product. This is when the product, design and architecture requirements must be fully aligned to be communicated efficiently to the engineering team so that their work can be planned accordingly and properly.

  As much as companies want it to be true, the overall process of getting something to market can never be 100% agile, the methodology preferred by a lot of technology firms. The first couple of steps of the process and the last couple are very linear and to an extent, similar to waterfall methodology. You cannot avoid it (as much as you want). One of these last linear steps is the Architecture review of the product requirements. A strong architecture team is one of the key elements of successful companies. While the natural approach would be to have former engineers with excellent communication skills be architects, I have noticed that the great architects may have had an engineering degree to start with but never worked in an engineering team. They were mostly coming straight from product and customer support.

The importance of formulating the "why" as a single item and the few "what" as a simple list will greatly enhance the speed and validation of the architecture review. The architecture team's role is to convert the product requirements into engineering requirements. Your role at that point is to review them all and make sure it will deliver the product you expect to deliver. Remember, a lot can be lost in translation. And all upcoming sprints will be based on the comprehension of the product requirements by the architect and how these were described to the engineer.

These "grooming" sessions are extremely important to pick up potential misunderstandings or problematic areas that could affect the timeline.

We were working on a project once, and everything was going as planned. We had skipped a couple of engineering demo days because the sprint they were working on was large, and it was better to see the bigger delivery instead of incomplete tiny bits left and right. When the day came to finally see how it looked, we were floored. It was terrible. One engineer had coded against a requirement precisely from a product viewpoint but left the engineers with three or four options on how to implement it. Instead of consulting with the architecture board on which one was the intended solution, he had picked the wrong one and went for it beyond the requirement because he "enjoyed" the challenge. The problem was this component of the solution was picked by other teams who also were confused at the beginning because it looked slightly different than the original requirement they got but didn't question it because they assumed it had been vetted by product and architecture and therefore, it was an approved change. In doubt, always ask if there was a change order approved is the lesson here.

Going back to grooming sessions, imagine these as basic troubleshooting. Even if the main goal is to align all requirements, a delivery person uses them to spot potential issues or misalignment. Listen carefully to what engineers may say as they "interpret" what was asked of them. Attending all these sessions and actively clarifying potential drawbacks or deciding the path forward when facing an issue is the key to successfully delivering. If you cannot be decisive on your project, don't expect others to do it for you and deliver the expected results. You will be set for disappointment. The main goal at the end of the building is to have a clear list of all the "how" against each "what." Building a roadmap with Epics and Stories is nothing more than that.

Congratulations, you have the engineering requirements done.

- *Testing*
The stepchild of many projects and yet the one that will make or break the delivery and the product's success. Because delivering is one thing, you must ensure it is usable; otherwise, you are not a delivery person but a garbage creator. It is a bit rough, but I want to drill into your mind the importance of proper testing to achieve a level of quality you can be proud of. Finding great quality assurance/control people is hard. Making sure they stay motivated and are treated with respect is harder. Many people see them as either failed engineers or project blockers when it is the opposite.

As for documentation, you want to test it at the beginning of any project. They should sit in the room during design and building. But if you can involve them during your analysis phase, you will realize how much time they will save you. Because they will understand the project's original purpose and feel some ownership in it, their goal will be to find and fix the problems before they become too complex.

When we were at PayPal working on mobile projects, we had two teams looking into testing the solutions we put out. One was a more senior team from an experience viewpoint, ensuring the integration with the payment platform and risk engine at PayPal was done properly. The other was a newly created mobile team with people understanding both the devices and the application. This led to interesting and challenging discussions during demo days and review sessions when a specific feature had failed one of the testing teams but passed with flying colors the other because they did not communicate their findings ahead of the review. The leaders of both teams worked out a system that allowed them to be in full coordination and the program manager collected bugs findings way before any group meetings. It was efficient to the maximum. On a funny side note, both remain good friends, with one leading a successful career at Amazon and recently retired while the other worked for a time with me at another company before launching a few start-ups of his own. So at least we knew that when we saw them walking together towards our working area, we were in trouble except around lunchtime, but it was a fifty-fifty toss.

Congratulations, you have the testing requirements done.

- *Deploying*
  Now that all the "'i" have been dotted and all the "t's" have been crossed in your project, it's time to deploy the solution. Whether it is a new product, service, platform or upgrade to an existing product already in the market, it does not make a difference. Too often, the executive team arbitrarily set the "launch" date to please the shareholders and Wall Street. This is not the way to go. I mentioned having a close relationship with the executive

champion and collecting information during casual conversations with other executives. When the date is given to you, and if it has not been communicated to investors publicly, you have some margin of negotiation. The best way to approach it is to go by acknowledging the request from the executive, reassuring them you will do everything possible to meet that date, but you want to double-check a couple of things before a hard commitment. After completing the analysis and the first round of discussion with other team members, it should give you some clarity on the feasibility or not of the date. The next challenge is to make sure the calendar and resources are aligned. Being part of one team is easy because everyone will work towards that delivery. But if you are working with other resources borrowed from other teams, you will have to compete for their time and availability hence why it is important to have the executive sponsor close by. Never call her/his name in vain. No one likes name-dropping in a company too much, but sometimes, it is good to remind them where the request you are delivering as a messenger came from. I said to be used rarely and wisely because executives like employees come and go. What may be the darling project of the day may be the financial burden of tomorrow. Play your cards carefully. That silver bullet is not to be wasted.

Making sure you remain in control of the "when" will be the part that will keep you up at night. I like to say it is like a giant three-dimensional game of Tetris; you have to constantly re-order based on variables you may not control but with a fixed time on the clock.

The importance of giving your team respect and trust will always be rewarded when deploying. While engineering has the luxury of working in different environments for development and production, testing is done in development only. So a lot is at stake when it is time to deploy into a "live" production environment. This is when you will need all hands on board. If testing is done properly and the infrastructure/IT team has been informed of it, it should go smoothly. And yes, we had a deployment once when the IT team decided to deploy a major update that same weekend because the person coordinating on their side left and forgot to inform them of that launch. We wasted seven hours trying to figure out why we could not access the elements we needed to deploy.

However, while everyone hopes and anticipates that deployment will go fine, you must also be ready mentally to go into a dark place. This is the one of a rollback where after giving it a go, you deploy in production to realize something is not working properly, and there is no easy fix. Most of the time, it will be database corruption or server failure.

We had a launch of a new web solution at a company. The engineering and IT team were based in India, with a small core support team in the US. The deployment was scheduled for a Sunday mid-day their time, a Saturday evening my time. I was the lead on the overall project, and since I had asked a few people on the team to be on call here in the US, I figured I should lead by example and was also available when it started. It was midnight when the deployment started. The first couple of hours went with no major issues, so I rested. Then we rolled in customer records, and at 2.20 AM, I got the call no one wanted to hear. The head of engineering was panicking, trying to figure out why the deployment was failing and so many components were corrupted. So I jumped in my car and drove to the office to fully view our protected infrastructure. I always remember the time it took me to find the main light switch in my area. From the outside, you had this guy walking around using his phone flashlight like a robber. It was especially nerve-wracking since burglaries had happened in that area the month before, and security was reinforced. Nothing happened, and I found the light. Going back to the deployment, we looked at what was going on frantically, trying to isolate the problem. Remember when I said to treat the testing team well? This is why. The QA lead called in 30% of her team on a Sunday to come and help because she had a vested interest in making the project succeed. We had been involved with her from day one. She knew most aspects of the project as we did. Despite this Dantesque effort, we could not isolate a single root cause. It was multiple elements that were failing. And this is what you have to be ready to go. Despite the "ask" to deliver the product before a major event in the industry, we would not be able to deploy on that day. I made the call. We did a rollback at 6 AM. This is the hardest call to make. Because you are not a delivery person anymore, you are the potential gravedigger of the project. I called my direct boss, who was the executive sponsor, at 8 AM, two hours before the planned call and informed him of the failed launch. The first round of questions was not about why I was calling so early on a Sunday morning but what its potential causes were. I had a long list. We went over it, and he agreed it was the right call and that not much could have been done differently. We had to figure out and see if we could salvage the marketing campaign due to start at the event three days later. And we salvaged it. The team was so frustrated not to have launched after so much anticipation that despite the massive lack of sleep for most of us, we troubleshooted every single aspect of it, and by that same Sunday evening Pacific time, which was then Monday morning India time, we had solved all issues. With some engineers on for more than 24 hours (not everyone can be Tom the Beast), we had a break to see if we could launch again on Monday evening India time. Going back through the SSOCCADD motion, we had to make sure we would not disrupt our major customers' activities as it was during the week (granted a Monday morning for the US, but we had customers all over the world) and that the infrastructure team was also on board and not doing any unplanned maintenance. We got the go-ahead, and launch number two was a success. We had made a date picked by the executive and marketing teams, but it involved a massive amount of stress.

Congratulations, you have a product out.

- *Maintaining*
  This is my least favorite part of any project. I like to create things from scratch or turn around and fix broken things. If it is working fine and humming at life, I get bored. And yet, it is important to plan the maintenance into the cycle of delivering a product. Both in the short term and long term but also the maintenance as a product and a piece of technology. Too often, when a project is launched and everything seems fine, resources are shifted to new projects, and potential problems are not anticipated. Maintain a minimum core team for key projects or products because you never know when something major will happen. And part of the maintenance is also to closely monitor the interaction of users with the product (it is delivering as designed or are there clear issues) and with the customer/technical support, which will collect a lot of precious feedback, especially the

negative one pointing to minor or major flows in the product (design or technology).

By maintaining a minimum viable team in place, you can also assure that the product runtime will remain in place. Plan to scale a product if successful, but sometimes things can go out of hand quickly. After the exhilaration of the successful rollout, when we launched Zelle, and while we thought we would catch our breath after months of insanely hard work, things went out of control. We had a great program manager, and she had anticipated this scenario as a strong potential. We always looked at it as a corner case, but it became a big-time challenge. Zelle became so successful that we were onboarding one hundred thousand new users daily. Yep, you read that right. Now you can understand that from a capacity management viewpoint, we had to scale fast, very fast. But because we had a scenario in place for that (even if we didn't believe we would have needed it in the first place), the team reacted promptly and put all the corrective measures in place, added capacity to the service and infrastructure, and we sustained these numbers for a few weeks.

And if the product has a longer life expectancy than some of the technical elements building it, it will be important to have people able to upgrade the technology along with people making sure it still complies with regulations and standards. I wonder how many times I have heard the massive failure of an online product or service was due to someone not renewing a certificate of some sort. It is actually the first thing I ask people in my team when credentials or login are failing. Did we do an upgrade, and have we moved/renewed all the certificates? It is amazing how many times this did the trick.

Congratulations, you have a successful product.

- *Documenting*
  The nerve of the war. You can have the greatest product on Earth, but without great documentation, it will fail. And I am not

talking of poorly translated user manuals in English or the Scandinavian building instructions for furniture that would challenge any brain. I am talking about documentation for highly complex financial technology products. Trust me, "user error" will be a permanent feature even with the best documentation, so better start early and get it complete. The biggest challenge is to have documentation digestible by people unfamiliar with the product or its functions. So the first order of business is to ensure that whoever will write that documentation is not speaking the company's internal lingo. Second, engineers should never write documentation. Let professional technical writers do this. The documentation quality will only be better and adapted to all audiences. Not everyone speaks the internal lingo of an engineering team. So give exposure to the product documentation team with users. At PayPal, we had a usability lab and members from multiple teams observed how people interacted with the product. We did similar exposure with the Mastercard Hackathon series, which was a great testbed for new products and their associated documentation. Again everyone working on a project should be their best advocate and critic. So do not hesitate to call it out if the documentation is terrible. In the technology product world, especially for APIs, templates have been well defined for years, and developers are used to reading through these super-fast. Following best practices will make your life easier as you build the documentation along the product progress. And keep the formatting consistent. I have seen the case of a product being structured differently from any other products we had on a portal. After multiple back-and-forths, they pushed back on modifying their documentation to match ours (and we were in the same company). In the end, developers got confused and skipped their documentation not based on best practices and then contacted developer support asking how it worked. This was using a lot of staff bandwidth. We finally used that product as a featured product during one of our big events and invited the leadership of that team to attend to see how developers interacted with it. Every participant stopped by the support table within the first hour to ask us how it worked because the documentation was unclear. It solved the problem right there on the spot, and a month later, we replaced their documentation with a newer one fully formatted the way we asked them to do in the first place. And we got out of the pushback in that event by having a workshop ready for the developers to help them jumpstart their coding process without depending on the documentation.

Congratulations, you have a usable product.

### Key Takeaway:
Don't be distracted by white or loud noises. Or false flattery. Focus.

# CHAPTER 7
# Deliver, deliver, deliver.

*"Perfect Day" - Lou Reed*

The more you deliver, the better you get at it. It's obvious. But as you go along the journey, you will identify people who are also excellent delivery people. Sometimes they know it, and sometimes they don't.

I have been working with people over a few companies, and we delivered. Like each time. Big challenge. Boom. Done. Keep these people close as great colleagues and sometimes great friends because you have gone through the trenches of delivering.

And when called again to deliver on that next big innovative idea someone had, you can either bring them on board with you or at least ask them for advice on how to tackle the next challenge.

Discipline will help you deliver, but it is also a condition for any company to survive. And have people who can execute. Someone was sharing with me how a company she was helping some time ago in the 1980s (delivering has been a consistent problem). They were asked to work with a new software startup and document their product. The company was well funded, which wasn't common back then.

Their office in Cambridge, Massachusetts, was in a newly renovated, very expensive building. The interior was gorgeous -- very expensive office furniture, artworks, stocked kitchen (way before Silicon valley made that a norm). Most of the staff was based in NYC, and they flew in for a day or two each week as needed and stayed in hotels. The organization chart was prominently posted -- all EVPs, SVP, VPs, and maybe a manager or two. So many chefs in the kitchen but no cooks.

Ten consultants brainstormed a marketing strategy for a product that didn't even yet exist. After the first day, the person who told me the story predicted they would run out of money before creating a product. She predicted six months but they went belly up after four months. Burn rate is one thing but not having a proper delivery plan with a clear business purpose to execute against is a failure in waiting. Among all these senior executives, I wondered how many saw themselves as a great visionary but couldn't foresee the writing on the wall. There was no talent to deliver. While most of what we discussed in this book may seem cliche and common sense, it is interesting to see how often people forget these basic steps. So, you may be the boring person who talks about the process, having a plan in place, following a timeline, and hitting roadmap milestones, but you are the critical element of the delivery. You are the delivery person.

As we all know, a great idea is important. But you have to build it to sell it. Lots of lessons to learn from this.

Sometimes, being the delivery person can seem like an over fetch in your job function. You may be a single contributor with no team put in charge of a project or product delivery. This means working with borrowed resources from other teams and with little influence on their motivation, availability or skills. Very rarely, except for critical projects coming from the top will you be given a "carte blanche" to put together your perfect team. It would be nice, but it is not the reality of the business. That is if you do not work for Apple. Apple introduced the concept of Directly Responsible Individual or DRI. Gloria Lin wrote back in 2012 an article explaining how it works in Forbes. A former Apple employee also described how it was to be a DRI at Apple, especially when multiple teams were involved. Everyone knows the stories of when someone from the design team under Jony Ive walked into any meeting, and there was an immediate submission by the rest of the attendees to whatever that person wanted. There may be some exaggeration there, but barely. However, the interesting part was that as an anointed DRI, even if you were a single contributor, you had full authority to get the project, product, and portion of what was assigned to you to its completion. This meant even pulling resources to help. Of course, the DRI was also expected to have a full and careful plan with resource usage laid out over the project's timeline. It was frowned upon to bring more resources even to solve an unforeseen and urgent issue. Frowned but tolerated because the DRI was the person supposed to know the best what was needed for the delivery of the project. If the DRI had an engineering issue, it was OK to pull resources from cross-functional teams with a full spectrum coverage of the potential problematic area. This even meant bringing external parties to help if needed.

So the DRI is the ultimate decision maker and the champion of their project. For Apple, it means there is no doubt about who is in charge. Have a question? Ask the DRI. Have an idea? Share with the DRI. Found an issue? Report to the DRI. The role and functions of the DRI is clearly a project manager on steroids because the enforcement part of the role is included with it. It also means that as a DRI, you must have a good grasp on every single aspect of the project because you may be the roster in the coop, but if things go wrong, it is your neck to be choked, and you become the feather in the duster to clean up the mess of a failed project. So yes, some DRIs may be difficult to work with, but it is their baby, and they are the ones losing sleep over issues, delays, conflicting inputs, bad metrics, etc. because they own it. And everyone wants their child to be the smartest, cutest, bestest in the world.

As you may find yourself a product manager, project manager, program manager or even executive sponsor of the next big thing, it is tempting to want to do only part of it. You may not be an expert in product marketing, corporate communication, regulatory compliance, coding, any functions part of a regular product cycle. I get it; it is scary to put your career up there on a job you may not even know if it is possible to build and deliver. You may be in the first few years of your professional life. You may be worried about not having the right experience or not being the most qualified person for the job. Well, it is normal. We all have self-doubt when facing a new challenge. And yet, as the saying goes, *"an entrepreneur is someone who jumps off a cliff and builds a plane on the way down."* It does not say if the plane flies. And that is why delivery people are important. The outcome is to be able to fly. And you can see that. The entrepreneur or innovator may just ask for a plane or something that looks like it. And that is your secret superpower. You know the entrepreneur doesn't want to die. You know he wants to safely land the plane after jumping from the cliff. You know there is little time to build a flying machine from when the feet leave the cliff to when there is an impact on the ground below. You know it doesn't have to be pretty as long as it flies. Do you know overall what a plane looks like? You know you may have to ask an aeronautics engineer to help you with the dynamics of flying, maybe a pilot to show you how to fly, an engineer to calculate the velocity needed, etc.

These you can read and anticipate in the first request. Surprisingly, this is not a skill everyone has. You can either have it naturally or work at it over time with multiple projects delivered to the world. For me, I am unsure which one is which. I always had a knack for finding the little details out of place. When situating under the SSOCCADD training, I would pick up the not weird but out-of-place element in my surroundings. So with any request coming my way, I always look at details others may have overlooked. When I was an intern many moons ago at British Airways, my boss could pick a numerical error in the middle of an Excel spreadsheet with thousands of lines and columns. I asked him how he did it. His answer changed the way I look at this document after. He said he knew what was supposed to be the total numbers at the end of each row and column. Not exactly, but in which ballpark because he had looked at these so many times over the years that instinctively, he knew what it should be. Then if he saw something that didn't match the pattern or was out of the expected range, he worked his way back up into the rows and columns until he saw it. He said he looked at a square block of about ten rows and ten columns at once. It helped him go faster.

In his mind, the delivered product was already built, and he was just checking the end-to-end process to ensure no stones were left unturned. A rare talent at the speed he was operating. And yes, these numbers may have cost you an increase in that ticket you were trying to book.

While the list of products and projects I delivered over the past thirty years of my active life may look impressive and organized gradually, there has been a lot of blood, sweat and pain in delivering some. It is one thing to make it look easy and to have the time to reflect on it calmly after release; it is another to have to make a hot analysis in the middle of a delivery. There will be arguments. There will be people quitting. There will be unforeseen technical or operational challenges. Yet, you must persist in getting the delivery done (especially if you are getting a paycheck for it). The size of the company you are delivering for is irrelevant. I have always delivered with a reasonably sized team. You don't need hundreds of people to succeed.

The key learning I got from participating in over fifty hackathons (as an organizer or attendee) was observing hackathon sharks. They are usually a premade team who comes in and goes for the win. You can make a good living while touring the globe by being a hackathon shark.

First, they know who they are coding for. They will check the judges before deciding whether they want to attend. Then they look at the reward of delivering.

Is it $10,000, $50,000, more? They are not in to find a job or to impress anyone. They are in for the money.

Secondly, after studying the technology (mostly an API or an SDK), they will learn as much about it before the event to hit the ground running (when you have 24h, every minute counts). If there is a theme, they will come up with an idea already on hand. Most professionally organized hackathons will check machines for recycled code as it is a big no-no. You must generate fresh code during the event.

Last, they are not going for perfection. They are going for a working prototype as in minimum viable product (MVP). This means they will narrow down to the bare core features needed to impress the judge. And then they code against achieving that goal.

The other important part of hackathon sharks is that they structure their team based on the skills needed to win. Most hackathons limit the size of teams to five members no more. Every shark team I observed was the same when composed of five people.

- The first is a person with a strong business sense, market understanding capacity, and a great pitch person. It is the product brain and storytelling voice for the team.
- Then you have the UI/UX designer. Most of the time, a great graphics artist comes from the gaming industry. They can turn around visuals (or fake them) to vow the judges to see way more than what's under the hood in reality.
- The third key player is a strong architect who can fuse all the product and technology requirements into a cohesive delivery and who sometimes knows and is still willing to code a little bit but, more importantly, will run the QA/Testing phase if needed (sometimes, time does not permit any testing :))
- The fourth one is your inevitable front-end developer. The one that will generate the web or mobile apps will include the elements to make it look real. They are the person you will find in the zone at 2 or 3 AM, sleep-deprived and fully caffeinated. Do not approach them when they are in the zone. You may lose a limb or two.

- And the last one is the Swiss army knife coder (or full stack if you prefer). Most of the time, a multi-talent engineer who can jump into front-end or back-end coding, make sure the underlying infrastructure is in place, solve coding errors on the fly and create the glue necessary to hold all the pieces together until the time to pitch the project to the judges. It doesn't have to be pretty, just to work.

These five profiles are mostly the perfect team you need to deliver, whether in a start-up or large organization. Find the talents, align the right skills, share the vision and hack the code.

### Key Takeaway:
There is a methodology to the surrounding madness. The good news is you don't have to find out alone.

# CHAPTER 8
# There is little glory.

*"Song 2" – Blur*

The description of the role of Directly Responsible Individual at Apple shows you are in the driver's seat when you are the delivery person, but you are also the only neck to choke if something goes wrong. Rarely the innovator or visionary will be "punished" for thinking boldly. So you will sometimes feel very lonely, which is even harder when you are not at the top. And as you are nose down trying to solve the latest challenge thrown at you, you will see the innovator glowing in the spotlight.

While the visionary will be most of the time in front of the media telling how great this new thing is, evangelizing the masses about how this is going to solve whatever problem it is solving, you, the delivery person, will be up at 11 PM on the weekend hoping a roll-back won't be necessary (when you deploy a technology and if it fails, you go back to the previous version).

When the visionary is doing a media interview at a big conference for your industry, you are collecting feedback from developers at 3 AM during a hackathon at the same conference.

So yes, there is little glamor and almost no glory. On top, you have to lead by example and take the hits for the team when it is necessary. No glory doesn't mean no sacrifice.

But despite getting most of the time secondary recognition, it will get you to places around the world and get you interesting gigs. When we were working on PayPal Mobile, we had the chance to travel to global locations to meet with other members of the team, discovering new countries for me, to attend conferences in fancy places (like Hawaii), and get nice rewards when patents are filed but more important, gaining an experience with little recognition but high values in many industries.

Even if you are working for a small start-up, investors are keen to get a product/service delivered by the companies in their portfolio. I developed close relationships with some investors in some companies I worked for. The information they were getting from the CEOs was OK, but they wanted to know how it was going in the trenches. Most of the time to make sure their investment may succeed so with a financial interest before the well-being of the team, but for some, it was genuinely because it allowed them to see the product evolving along the delivery path while also reflecting on the market need. Most venture capitalists are used to hearing pitches insanely not connected to reality with overinflated market analysis and over-aggressive delivery timelines and features. When I was asking a few how they get the patience to listen to these, the most common answer is that they know they are being BSed and that whatever is on the financial slide, including market projection, is good for the trash after the presentation or the funding because it is unrealistic. And yet, sometimes, they make the jump and invest. And most of the time, it is not for what is being pitched to them but because what they see in the team presenting is the capacity to deliver even if it is not on that round of investment. The "bankability" of the team is critical.

Even Silicon Valley venture capitalists understand the importance of having the right set of talents to deliver and to make sure they can pivot and adapt to the circumstances while still following the main plan.

One aspect of having delivered some impactful projects and worked for well-known companies is the access to many entrepreneurs and people with ideas. The positive of being a delivery person is that you can apply this to something other than a product or service.

I have helped countless start-ups put together content for venture capitalists and taught them how to pitch. There is little difference from what we talked about in this book. You have someone with a vision but who doesn't know how to pitch or even structure a deck to present to investors. This is when the focus and capacity to see the endgame come in handy. There was a start-up I was helping some time ago. Their product was advanced and could change how skin cancer treatment would have been approached. The original idea came from this genius scientist out of NASA. He had put together a small team around him and a "specialist" in funding. That the founder is a bit absent-minded professor was expected, and he was not. He was very driven and focused but raising money was not his cup of tea, so he wanted help. I met with his "finance specialist" and was floored. Despite claiming a lot of experience, that person had never pitched (successfully) to Silicon Valley sharks, sorry Vulture Capitalist, as Jean-Louis Gassée would say.

Here is one of the well-known secrets of Silicon Valley. Venture Capitalists have a very short attention span. So when presenting anything to them, it needs to be short. The presentation this person had created was forty-two slides long. And even with my color vision issues, I was blinded by the purples, violets and other super aggressive color combos that would have given a stomach ulcer to a chameleon. So the CEO and I sat down to rework the pitch deck for the meeting we had two days later.

First, I told them we had to get from forty-two slides to three. Here is another secret, you need three slides to convince an investor. What are these? Here you go

- Slide #1 should present the market need, size and, more importantly, the problem being solved. This is your product. If the problem you are solving comes from a personal experience that gave you the idea of how to solve it, you get bonus points.

- Slide #2 is the presentation of the team. Who are you? What have you done and where? For how long have you known each other? Have you worked together before you started this new start-up? This is where the venture capitalist checks the chemistry of the team and the commitment to each other to succeed.

- Slide #3 highlights the financials or how you and the investors will make money together, but more importantly, how the investors will make more money than you. Just kidding, but this is what Jean-Louis Gassée describes as the

Money Pump. Have sharp numbers grounded into factual market analysis, not a moonshot. Every venture capitalist knows you cannot have exactitude at that stage, but they have seen so many pitches they can compare to previous financial forecasts to see if they can be confirmed as realistic or not. So be ambitious but be "reasonable," or your bluff will be called out, and the investor will take a hard pass.

As we worked on these slides, making them more blended from a color viewpoint (clean white presentations were fashionable again at the time with Apple going through its super simple white phase for everything), we arrived at the magic three slides. We kept only a dozen as a backup to cover potential questions. It is OK to have backup slides as it shows you have anticipated the questions from the investors (which they will appreciate), and you were mindful of their time by not going for death by PowerPoint.

We practiced for a few hours the day before the official pitch and made sure the CEO was comfortable and the "funding" expert was ready as a backup to work on questions if the term sheets discussion happened. The day of the meeting arrived, and I was waiting outside the office of the first venture capitalist firms we were meeting that day in Palo Alto and Sand Hill. The funding expert showed up alone. I asked him where the CEO was, and he told me his flights had been delayed and he would not be able to make it, so it was up to us. We walked into the conference room, acknowledging the firm's two partners (one of whom was an acquaintance from a previous company I worked for and into which he had invested). And then, everything went downhill. Mister Funding Expert pulled his purple presentation with forty-two slides and mumbled a narrative that was more dispersed than a Fourth of July firework show. Despite my multiple attempts to reset the meeting and go back to the original plan, he dismissed any dialog, ignoring me (as I was a temporary consultant anyway) and kept going through his slides. He didn't go well. After slide #8, if I recall correctly and twenty minutes down the drain, one of the partners stood up and said, "What is this about? I don't get it, and you guys don't seem ready for pitching yet.

Come back when you are." And he left. It was the end of the meeting as their other partner, the younger one I knew, shook his head and gave me an underside look with sadness for him and me. Not the greatest feeling on Earth.

Five minutes later, we were on the sidewalk outside their office. We had agreed to call the CEO and report to him the takeaways. I was fuming. I warned them the day before that we had a limited number of meetings we could get with venture firms in Silicon Valley or the six bullets. If six firms tell you "no," don't even try to go to the seventh one. They talk together, and if you shot your six bullets and hit nothing, your idea is dead on its track. I feel we had wasted a bullet because we didn't stick to the plan. I was ready to share my feelings with the CEO, but our "funding" expert said something which, to this day, still puzzled me. In his view, the meeting went very well, he presented the "original" better presentation, and the investors had great questions and invited us back to meet with them. He had completely missed the point. They didn't invite us back. It was a polite way to get us out of the door without even having the reason to make a yeah or nay choice. It was a cold pass without a need for further explanation. I was floored.

The CEO was pleased and said it was encouraging for the afternoon meeting. Since we were using the phone of the "expert," I told him on speaker phone I would need to call him from my phone before. We went for lunch on University Avenue in Palo Alto, and while the "expert" was browsing the menu, I stepped outside and called the CEO. I told him the meeting was catastrophic and that we wasted the time of these two people in the morning, including a friend who now looked bad to his partner, and that if we couldn't stick to a simple plan, I would not go to that afternoon meeting. He tried to calm me down and said he had talked with the "funding" expert in the morning, and he had convinced him the original slides were better at browsing a better picture of the product. I told him it was not the point of the meeting. We were not here to educate the investors in becoming experts on the topic but to have enough confidence we were the right experts to give us money to pursue the project. We had wasted a bullet; if the plan persisted down that path, I was out.

Meanwhile, I also asked him why he was not there since his plane should have landed by now. He said his flight had been canceled, he could not make it to the meetings we had planned for that day, and he was unsure about the day after. So here we were with the CEO and founder of the company not being around and a pseudo-expert sabotaging the chance to get term sheets faster than a venture capitalist could say no (and they are fast). So I told him I was out and could not help them further at that moment as it seemed my expertise was not needed. Later I learnt that he never intended to come to the meetings because he did not like financial discussions and was worried his idea would be seen negatively by investors, and he didn't want to see it. The work I did for weeks simplifying the project, presenting it in the proper format and making sure we had the right endorsement and audiences (venture funds are specialized in specific market segments, so pitch only to the relevant ones and never carpet bomb your executive summary to investors who will never invest in your domain).

Later, we pitched to another friendly investor using the correct slide. There was interest, and it went to deeper discussion, due diligence, etc., but not investment. I had taken a cold shower with them before, so I didn't have the same fire in my belly anymore. Maybe my friend at the Venture firm noticed it, or maybe it was just time to have the product out. It happens all the time.

The important part here is that you can use the VC presentation approach to give yourself a high-level view of what you want to deliver. Do you understand the "what, who and how" and of course, the time and budget it will take? It's that simple sometimes to remain focused.

You can see that the skills of a delivery person can be applied to many aspects of professional interests. Follow the process, rinse and repeat if you do what you like and stick to your guns when you need to.

The active osmosis between the innovator and the delivery person is not just limited to this duo. You will discover that other factors also affect the way you will plan. I mentioned earlier how it was important to not commit to a delivery date if it was unreasonable. Unfortunately, in corporate America, decisions are sometimes made without considering the situation on the ground. The Event Marketing team may want to do a splash at an event by announcing something big. The Investor Relations team may want to dilute bad financial news by announcing something new. The Executive team may want to close a partnership or acquisition deal but need to show the technology is there. It can be anything. And when you are not part of the decision, trying to make it work is very lonely.

The Twittersphere provided one of the best explanations. When working on a delivery planning phase, you should always have a Data-Driven Decision Making, but unfortunately, it is most of the time a Decision Driven Data Making. The decision has been made, and you have to find data to support it. You will learn to live with the shame. Remember that you must be realistic and transparent when you have put together the delivery plan. Even if the decision was made, you can always go back and ask for more justifications from the Market Analysis team or the Product Team, etc., about the whys. Granted, it is easier said than done, but if you don't ask, you don't know.

While you will encounter countless challenges, and it will seem that sometimes the reward is not worth the sacrifices, see it as a temporary setback to set you up for your next more fun project. Just don't be miserable at it. Or try to get some of that visionary pixy dust by asking the marketing team if you can have a speaker spot at an event or get to write an article. Even the delivery person doesn't mind some recognition sometimes, and you are the best ambassador of your brand.

### Key Takeaway:
It's hard work. There will be discouraging and lonely moments, but the overall journey is what you make out of it. Create your own fun.

# Delivery Words

*"The end" – The Doors*

Hopefully, your takeaways from this book are more than just business cliches but also some wisdom acquired via the trials and errors of my career. You have the tools with the SSOCCADD approach, know who you need to include in your team to succeed, have the support you need from the top, and have the energy to drive the project to the end.

Having the luck to work for a great visionary or leader will make your career shine in a way that cannot be acquired anywhere. Sometimes, you may be in the shadow of greatness and don't feel bad about it. It is a great spot to observe and learn.

While observing, try to learn from everyone around you. Try to recognize people you admire, their strengths/skills, and some recurring traits you find common among them. I had great bosses I am still in touch with today. The most common trait was humility. Despite their role as top executive, they were not using their power because they didn't need to. They are leaders, not bosses. The first one to believe in me was Harald L. When the team was let go from Motorola, he found a new role for himself but negotiated to bring his core team to the new company. It was rewarding to know someone you admire was looking out for you and willing to trust you again.

Each leader I admire has remained in contact, and we sometimes connect on work-related questions or personnel matters. Even when retired or overseas, they will make time to chat, which is priceless help. Eric D., Laura C., Mario S. or Brandon R are part of a sounding board I call upon when in doubt.

So as one last piece of advice. How I make my life simpler as a delivery person. In my first few roles as a people manager, when I was at about the middle point of my career, I had to make tough decisions, but I was sure who to ask for advice and guidance. I had friends I could go to, family, and acquaintances, but I wanted continuity in the support so I would not have to re-explain or go back to some past decisions with a new person each time.

While it may be obvious to have people you can trust, I created my personal board of directors. I shared with some the concept, and it didn't seem to shock them, so I kept it going.

The board's composition has been constant over the years, with very few changes.

- *Chairman of the Board:* Of course, it is my wife, Rebekah. She has a 360° view of my life, personal and professional. She knows how my brain functions, and there is no need for a long explanation.

- *Technology Legend:* If you are lucky to know someone who is a top influencer in the same professional world as yours, ask them. For me, Jean-Louis Gassée is that person. Despite seeing me evolving from an acned teen in shorts to who I am today, he still kindly and patiently gives me some of his precious time to go over my existential questions and problems to solve. A lot of the nuggets of wisdom in this book come from him, actually.

- *Former Boss:* As I remain in touch with former bosses I liked, there is always one I will get more often than the others. Sorry, guys, no jealousy here. But it doesn't mean I will not alternate who I am reaching out to based on the question I am trying to get answers to. The nice part is we can have open and freer discussions than when there was a reporting line. Some of the best career advice came from these post-work discussions when some important pieces of behavioral observation could be shared without the risk of an HR review or a permanent mark on an annual review.

- *Financial Expert/Trust Source:* For all decisions related to a job change, a promotion, a demotion, an industry change, etc., there will be a financial impact. It is always good to have someone who can bring expert views on all potential gains and losses linked to a decision. I am lucky enough to have that person for many years in the person of my uncle Serge Yablonsky, who, in his roles as Chartered Accountant and Financial Technology Consultant, has moved in similar spheres of work expertise to me. And him being family, we can have a very frank and open discussion. He is not shy to let me know when I am about to do something he would qualify as "damageable". He is also the reason I work in the technology field. We share more than blood. We share the same love for Mac. In the 1980s, when Apple France was just starting, I remember seeing photos of my uncle and Jean-Louis Gassée in magazines sitting next to beautiful devices. And then my uncle sat me down in front of one of them at his home and told me to just play with it, explore it. I could have done a lot of damage to it, but it was not his concern. When I discovered MacPaint, I spent hours and hours drawing and creating. It also saved me time sitting at the family dinner table listening to adult conversations. I could escape into this new computing world, and the adults were approving. I wanted to be a computer graphic artist. It didn't work that way but close enough. He was also the one who set me up with my first internship in the USA. My aunt and him paid for my airfare to come to America. We always joke it was a one-way ticket, but it was really an open return ticket that I haven't used in more than 25 years. And with TWA being gone, I am unsure of its validity.

- *Friend (s)*: Lots of former colleagues became friends. And friends from outside of work. Depending on what you are trying to decide and if you are lucky enough to have a stable circle of friends (Silicon Valley can be tough for this), you can try to have 2 or 3 close friends to whom you can ask for their opinions when facing a choice to be made. Some of my Googlers and former PayPalian friends have been cornered by me few times when trying to find a solution for a work problem or a project being blocked. Brainstorming in a friendly, candid way (and sometimes with a few drinks) can be stimulating and rewarding. Usually, the most obvious solution may not be the best or most impactful one. At least as friends, they should always have your best interest in mind.

The delivery man can be a lonely role, but it doesn't mean it must be done alone. Always try to use all resources available to you to keep simplifying the process, the work politics, the meaning of life, etc. and never be shy to ask for help. More people will help than you may think is possible.

And when that big project is done, the product is in the wide, wild world, and you see people enjoying using it, stay humble and reward everyone who helped you to get there. Remember the great but also dark moments. Learn from these as it will help you grow and create automatism on solving issues in the future. Even if you have done it many times, there is always room to grow and learn. The more you add to your arsenal of knowledge, the better and faster you will get at delivering. You may not have the biggest innovative idea, but you can dictate the circumstances of how it will become a reality. You are the delivery person!

Every great creator or dreamer needs a sidekick to help them. In the 17th century, Miguel de Cervantes wrote Don Quixote. Honestly, I made most of my living across all these years by being Sancho Panza to many Don Quixotes. And I have zero regrets. The adventures were great.

Now, I am entering this phase in my career where transmitting knowledge, mentoring and helping others find their true calling is as important as delivering. So reach out and share your experience. With all the new technologies coming, Web3.0, Blockchain, DeFi, etc. the process of delivering will become even more complex and decentralized. This is a new frontier to explore for all of us and it is never too late to learn more even for Sancho, the Delivery Man.

## *Thank you.*

I want to extend my thanks to all my former colleagues who were in the trenches with me in the wee hours of many nights, trying to pull what felt like miracles sometimes. To my great bosses, Harald L.(RIP), Eric D., Laura C., Brandon R., Mario S., Kenzo F.. You trusted me at a point in time and shaped me into becoming a better self.

Thanks to the great visionaries who allowed me to get some of their pixie dust, John Wood, David Marcus, Ajay Banga, Ann Cairns, Stan Chudnovsky and few others…

Thanks to Paula Berger, Aneace Haddad., Julia Collins, Marc Verstaen for providing me some extra confidence and example to make this book complete.

Thanks to Kate Niendorf for working on the SSOCCADD visuals with little guidance.

Thanks to Onur Burc for the cover design and his absolute patience.

Thanks to Sam Wright and the Yeti Studio crew who provided the needed editing and formatting to my sometimes long mumbling and transformed this book into a readable piece. Magical talents are real.

Thanks to my personal board and especially John Wood, Serge Yablonsky, Jean-Louis Gassée and David Birch. Your friendship, countless hours of free counseling, advising, guiding, and mentoring, even if you didn't know it, have been essential to giving me the courage to put words on paper. I don't know if it will help anyone, but at least it helped me as you have always helped me.

And last, I should and would not forget to thank my family. To my parents who still don't know what I do for a living and just hope it is honest work. To my kids Rachelle and Aleksandre for tolerating my long physical absences, my mood swings, my occasional emotional detachment when in deep thinking and for keeping me grounded. And to my wife Rebekah for your constant patience and support. I like to think that you are my delivery partner, even if you don't understand the weird world of Silicon Valley technologies, and I don't always get your world of teaching. Sometimes, it is just nice to be the husband of the doctor. You are my sanity rock.

# "About the Author"

Sebastien Taveau is a puzzle solver and beyond-the-horizon watcher.

Seb's technical and professional experience spans over 25 years in various industries. He has shaped ecosystems and products around FinTech, mobile payment, mobile security, mobile identity and consumer solutions even when he was told it was impossible.

He has provided expert opinions for and has been quoted in the WSJ, Washington Post, The Huffington Post, Reuters, Mashable, USA Today, CNN, CBC, Forbes, Dark Reading, Digital Transactions, Newsweek, etc. on topics ranging from mobile payments to mobile identity, and consumer biometrics and security. He is a keynote speaker and panelist to many industry events and guest lecturer at universities worldwide.

Most recent projects include Zelle (the real-time P2P) as Chief Technologist, Mastercard Masters of

Code (one of the most successful developer engagements in the world of payment) as Chief Developer Evangelist, Envestnet | Yodlee as Head of Developer Experience delivering top APIs, developer portals and other engagements with coders around Wealthtech, Human Interface technology for Android devices (via Validity) as Chief Technology Officer, PayPal Mobile among others where Sebastien led or was part of the core execution/strategy/executive teams.

Seb is also an inventor with more than 25 patents granted and was a member of the exclusive eBay Inventors.

Seb lives in the Bay Area where he spend most of his time with his wife and his two children. For mental resourcing, you will find Seb roaming in the French/Swiss Alps.

The event that started it all.
Explaining to the Banrisul Tech Summit the
concept of the 4 "i's"
(and few more like Inspiration, Idea, etc.)

Taking a short pause during a Mastercard
Masters of Code Event

Jean-Louis Gassée as keynote speaker for
Masters of Code SF

Jean-Louis Gassée and Adam Paul
(Actor/Director aka The Naked Man in
"How I Met Your Mother" TV show)

having fun at the Grand Finale of the
Masters of Code

Adam Paul, Dave Birch

Dave Birch and I in intense debate before a
panel

Dave Birch and I are trying to decide which
football club deserves the Cup.

Make big plans.

Finding inspiration all around you. Even in
the middle of a delivery, do not miss
small relief signs.

The joy of seeing crowds using your product
or coming to your event is
the ultimate reward of delivering.

Sometimes a little break (foosball is a must have in any developer world)

Being creative to keep delivering, like renting a massive generator / "mobile" servers in Brazil

With Harald L., one of my first bosses who mentored a lot of skills in my mind.

With Eric Ries, author of Lean Start-up

With Adam Paul and Ben Parr (Co-founder
of Octane AI, Author of Captivology and
Co-host of Business Envy - former
Co-Editor of Mashable)

With Michael Dell

With Professor Scott Galloway

With Karsten Nohl
One of the top grey hats in the world. We
used twice the talent of his team at Security
Research Labs in Berlin.

With John Collison,
Stripe co-founder

With Chris Hardwick the Nerdist

Made in the USA
Monee, IL
22 November 2022

316f0c77-be2a-4f8b-b449-5976a57bad47R01